GEOMETRY,
RELATIVITY and the
FOURTH DIMENSION

GEOMETRY, RELATIVITY and the FOURTH DIMENSION

by Rudolf v.B. Rucker

Department of Mathematics,
State University College of Arts and Science, Geneseo, N.Y.

Dover Publications, Inc., New York

Published in Canada by General Publishing Company, Ltd., 30 Lesmill Road, Don Mills, Toronto, Ontario.
Published in the United Kingdom by Constable and Company, Ltd., 10 Orange Street, London WC2H 7EG.

Geometry, Relativity and the Fourth Dimension is a new work, first published by Dover Publications, Inc., in 1977.

International Standard Book Number:
0-486-23400-2
Library of Congress Catalog Card Number:
76-22240

Manufactured in the United States of America
Dover Publications, Inc.
180 Varick Street
New York, N. Y. 10014

PREFACE

This book is about the fourth dimension and the structure of our universe. My goal has been to present an intuitive picture of the curved space-time we call home. There are a number of excellent introductions to the separate topics treated here, but there has been no prior weaving of them into a sustained visual account. I looked for a book like this for many years—and finding none, I wrote it.

Geometry, Relativity and the Fourth Dimension is written in the hope that any interested person can enjoy it. I would only advise the casual reader to be willing to skim through those few sections that may seem too purely mathematical. This book is, however, more than a standard popular exposition. There is a great deal of original material here, and even the experienced mathematician or physicist will find unexpected novelties.

I am indebted to all of the authors whose work is described in the Bibliography, but most especially to Edwin Abbott, Arthur Eddington, Hans Reichenbach and John Wheeler.

R. v. B. R.

Geneseo, N.Y.
January 31, 1976

94631

CONTENTS

GEOMETRY,
RELATIVITY and the
FOURTH DIMENSION

1

THE FOURTH DIMENSION

We live in three-dimensional space. That is, motion in our space has three degrees of freedom—no fewer and no more. In other words, we have three mutually perpendicular types of motion (left/right, forward/backward, up/down), and any point in our space can be reached by combining the three possible types of motion (e.g., "Walk straight ahead about 200 paces to the river, then go right about 50 paces until you come to a big oak tree. Climb about 40 feet up it. I'll be waiting for you there."). Normally it is difficult for us to perform up/down motions; space is more three-dimensional for a bird or a fish than it is for us. On the other hand, space is essentially one-dimensional for a car driving down a two-lane road, essentially two-dimensional for a snowmobile or a car driving around an empty parking lot.

How could there be a fourth dimension, a direction perpendicular to every direction that we can indicate in our three-dimensional space? In order to get a better understanding of what a "fourth dimension" might mean, consider the following sequence:

We take a 0-D point (Figure 1; from now on, we'll abbreviate "n-dimensional" by "n-D"), move the point one unit to the right (this produces a 1-D line segment, Figure 2), move this segment one unit downward (this, with the lines connecting the old and new segments, produces a 2-D square, Figure 3) and move the square one unit forward out of the paper to produce a 3-D cube (Figure 4).

Notice that we cannot actually draw a 3-D cube on this 2-D sheet of paper. We represent the third dimension by a line that is diagonal (rather than perpendicular) to the left/right and up/down dimensions. Now, we don't really know anything yet about the fourth

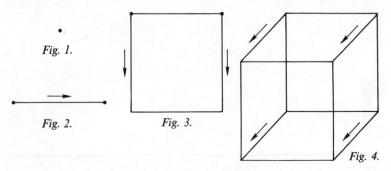

Fig. 1.

Fig. 2.

Fig. 3.

Fig. 4.

dimension, but couldn't we try representing it by a direction on the paper that is perpendicular to the (diagonal) direction we used to represent the third dimension?

If we do so, we can continue our sequence by moving the cube one unit in the direction of the fourth dimension, producing a 4-D *hypercube* (Figure 5).

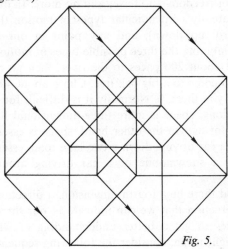

Fig. 5.

This design for the hypercube is taken from a little 1913 book, *A Primer of Higher Space*, by Claude Bragdon, an architect who incorporated this and other 4-D designs into such structures as the Rochester Chamber of Commerce Building.

It is also possible to consider a similar sequence of spheres of various dimensions. A sphere is given by its center and its radius; thus the sphere with center 0 and radius 1 is the set of all points P such that the distance between 0 and P is 1. This definition is independent of the number of dimensions your space has. There is no such thing as a 0-D sphere of radius 1, since a 0-D space has only one point. A 1-D sphere of radius 1 around 0 consists of two points (Figure 6).

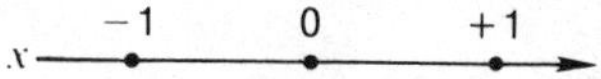

$$x \xrightarrow{\quad \overset{-1}{\bullet} \quad \overset{0}{\bullet} \quad \overset{+1}{\bullet} \quad} \qquad |x| = 1$$

Fig. 6.

A 2-D sphere of radius 1 can be represented by this figure in the *xy*-plane (Figure 7).

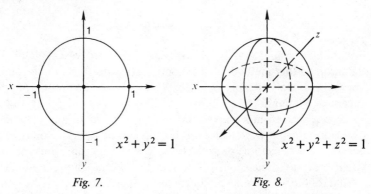

$$x^2 + y^2 = 1$$

Fig. 7.

$$x^2 + y^2 + z^2 = 1$$

Fig. 8.

A 3-D sphere of radius 1 in the *xyz* coordinate system looks like Figure 8.

Although, reasoning by analogy, a 4-D sphere (*hypersphere*) can be seen to be the set of quadruples (x, y, z, t) such that $x^2 + y^2 + z^2 + t^2 = 1$ in the *xyzt* coordinate system, we cannot say that we have a very good mental image of the hypersphere. Interestingly, mathematical analysis does not require an image, and we can actually use calculus to find out how much 4-D space is inside a hypersphere of a given radius r.

The 1-D space inside a 1-D sphere of radius r is the length $2r$.

The 2-D space inside a 2-D sphere of radius r is the area πr^2.

The 3-D space inside a 3-D sphere of radius r is the volume $4/3 \pi r^3$.

The 4-D space inside a 4-D sphere of radius r is the *hypervolume* $1/2 \pi^2 r^4$.

One of the most effective methods for imagining the fourth dimension is the method of analogy. That is, in trying to imagine how 4-D objects might appear to us, it is a great help to consider the analogous efforts of a 2-D being to imagine how 3-D objects might appear to him. The 2-D being whose efforts we will consider is named A. Square (Figure 9) and he lives in Flatland.

A. Square first appeared in the book *Flatland*, written by Edwin

Fig. 9.

A. Abbott around 1884. It is not clear if Abbott was actually the originator of this method of developing our intuition of the fourth dimension; Plato's allegory of the cave can be seen as prefiguring the concept of Flatland.

A. Square can move up/down or left/right or in any combination of these two types of motion, but he can never move out of the plane of this sheet of paper. He is completely oblivious of the existence of any dimensions other than the two he knows, and when A. Sphere shows up one night to turn A. Square on to the third dimension, he has a rough time.

The first thing A. Sphere tried was to simply move right through the space in A. Square's study. When A. Sphere first came into contact with the 2-D section of his 3-D space which was Flatland, A. Square saw a point (Figure 10). As A. Sphere continued his motion the point grew into a small circle (Figure 11). Which became larger (Figure 12). And then smaller (Figure 13). And finally shrank back to a point (Figure 14), which disappeared.

A. Square's interpretation of this strange apparition was, "He must be no Circle at all, but some extremely clever juggler." And what would you say if you heard a spectral voice proclaim, "I am A. Hypersphere. I would teach you of the fourth dimension, and to that end I will now pass through your space," and if you then saw a point appear which slowly inflated into a good-sized sphere which then shrank back to a point, which winked mockingly out of existence. We can compare A. Square's experience and yours by putting them in comic-strip form, one above the other (Figure 15).

The difference between the two experiences is that we can easily see how to stack the circles up in the third dimension so as to produce a sphere, but it is not at all clear how we are to stack the spheres up in the fourth dimension so as to produce a hypersphere (Figure 16).

We can, however, work out some possible suggestions. One is

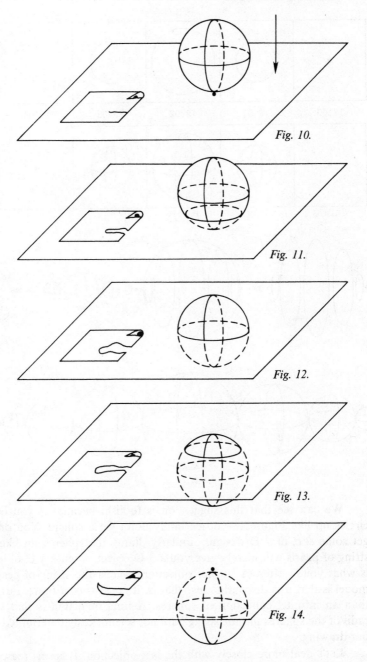

Fig. 10.

Fig. 11.

Fig. 12.

Fig. 13.

Fig. 14.

that the spheres might be just lined up like pearls on a string, and that a hypersphere looks like Figure 17.

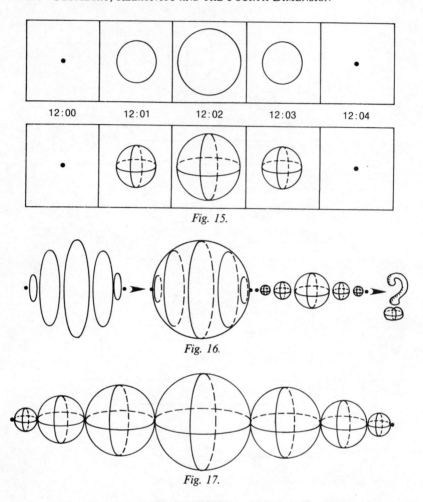

12:00   12:01   12:02   12:03   12:04

Fig. 15.

Fig. 16.

Fig. 17.

We can see that this suggestion is foolish, because if you line circles up like Figure 18 you certainly don't get a sphere. You only get some sort of 2-D design. Similarly, lining the spheres up like a string of pearls will merely give you a 3-D object, when a 4-D object is what you're after. A further objection against the string-of-pearls model is that it is discontinuous; that is, it consists of a finite, rather than an infinite, collection of spheres. A final objection is that the radii of the spheres in the "string" are not scientifically determined in our drawing.

Let's deal more closely with the last objection. It seems reasonable that the length of the "string" should be equal to the diameter of the largest sphere. The idea is that we will have a sphere moving

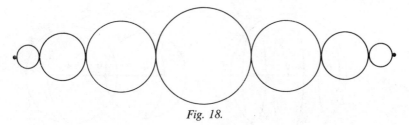

Fig. 18.

along this length, starting as a point, then expanding to the size of the largest sphere, and then contracting back to a point. To get the picture, let's talk for a minute in terms of turning a 3-D sphere into a 2-D figure. Imagine slicing a 3-D sphere up into infinitely many circles. Then imagine simultaneously rotating each of these circles around its vertical diameter through 90°. The sphere will thus be turned into a 2-D figure consisting of infinitely many overlapping circles. The process can be compared to what happens when you pull the string on a venetian blind to turn all the slats from a horizontal to a vertical position. The resulting 2-D figure looks like Figure 19.

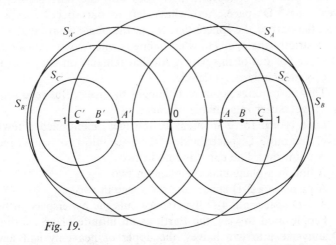

Fig. 19.

Notice that the radius of each of the component circles of this "closed venetian blind" version of the 3-D sphere is equal to the vertical distance between its center and S_0, the circle whose radius is the same as that of the 3-D sphere (Figure 20).

Now, if you take Figure 19 and replace each of its circles by a sphere, you get something that is a solid made up of infinitely many hollow 3-D spheres. Recall that the way in which we turned the 2-D figure (an area made up of infinitely many circles) into a 3-D sphere was by rotating each of its component circles 90° around its vertical

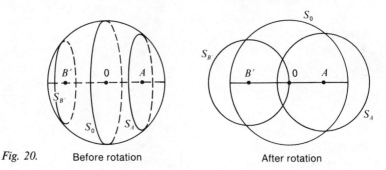

Fig. 20.   Before rotation   After rotation

axis. So it seems that the way to turn the 3-D solid that we have imagined into a 4-D hypersphere is to rotate each of its component spheres 90° around the plane that cuts the poles and is perpendicular to this sheet of paper. How do you rotate a sphere around a plane? As we'll see in a while, this isn't too hard if you can move through the fourth dimension. What's left of a sphere after you rotate it in this way? Well, half the sphere goes into the part of 4-D space "under" our 3-D space and half goes into the part of 4-D space "over" our 3-D space. And what's left in our space? Just a great circle, the part of the sphere that lay in the plane we rotated around. This is strictly analogous to what happens when you rotate a circle in 3-D space 90° out of this paper. All that remains on the paper is two points of the circle, a 1-D circle.

This all requires some real thought to digest. But read on, read on. It'll get easier in a couple of pages.

Let's return for a moment to the idea, mentioned a few lines above, that our 3-D space divides 4-D space into two distinct regions.

A point on a line cuts the line in two.

A line in a plane cuts the plane in two.

A plane in a 3-D space cuts the space in two.

A 3-D space in a 4-D hyperspace cuts the hyperspace in two.

People used to view the Earth as an infinite plane dividing the 3-D universe into two halves, the upper or heavenly half and the lower or infernal half. If we assume that the 3-D space we occupy is flat (in a sense that we will make clear in a later chapter), then we can conceive of Heaven and Hell as being two parts of 4-D space which are separated only by our 3-D universe. Any angel thrown out of Heaven has to pass through our space before he can get to Hell.

Now, if a hypersphere has been placed so that its intersection with our 3-D space is as large a 3-D sphere as possible, it will be cut into a heavenly hemihypersphere and a hellish hemihypersphere. We can use this idea to get a new way of imagining the hypersphere.

If you take a regular sphere and crush its northern and southern hemispheres into the plane of the equator, you get a disk, or solid circle. Similarly, we can imagine crushing the heavenly and infernal hemihyperspheres into the space of the hypersphere's largest component sphere to get a solid sphere. The solid sphere can be turned back into a hypersphere if we can somehow pull its insides in two directions perpendicular to all of our space directions. How do you do this? Well, how would you pull a solid circle out into a sphere? Imagine that the inner concentric circles belong, alternately, to the northern and the southern hemispheres. You can pull them in opposite directions without having them pass through each other (Figure 21). So to decollapse our solid sphere we pull its concentric spheres alternately heavenward and toward the infernal regions.

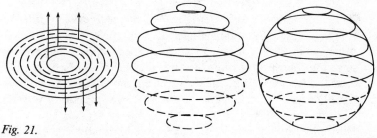

Fig. 21.

In this discussion of the hypersphere I've drawn on some new ideas about the fourth dimension: One is that you can rotate a 3-D object about a plane to leave only a plane cross section of this object in our space. Another is that you can "move through obstacles" without penetrating them, by passing in the direction of the fourth dimension. To clarify these, and other ideas, let's get back to good old A. Square.

After the sphere showed himself to A. Square, A. Square remained unconvinced. So A. Sphere did some more tricks. First he removed an object from a sealed chest in A. Square's room—without opening the chest and without breaking any of its walls. How was this possible? A chest in Flatland is just a closed 2-D figure, such as a rectangle (Figure 22). But we can reach in from the third dimension without breaking through the trunk's "walls" (Figure 23).

The analogy is that a 4-D creature should be able to, say, remove the yolk from an egg without breaking the shell, or take all the money out of a safe without opening the safe or passing through any of its walls, or appear in front of you in a closed room without coming through the door, walls, floor or ceiling. The idea is not that

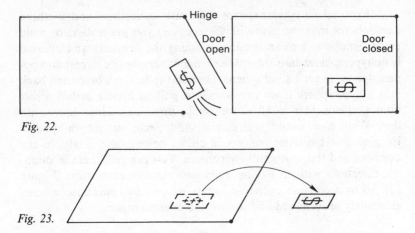

Fig. 22.

Fig. 23.

the 4-D being somehow "dematerializes" or ceases to exist in order to get through a closed door. Your finger does not have to cease to exist for an instant in order for you to put it inside a square. The idea is that since the fourth dimension is perpendicular to all of our normal 3-D space directions, our enclosures have no walls against this direction. Everything on Earth lies open to a 4-D spectator, even the inside of your heart.

The only way in which A. Sphere could finally convince A. Square of the reality of the third dimension was to actually lift him out of Flatland and *show* him what it was like to move in three dimensions. Is there any hope of this happening to us? Is it likely that there are 4-D beings who, if summoned by the proper sequence of actions, will lift us out of our cramped three dimensions and show us the "real stuff"? A lot of people used to think so at the time of the Spiritualist movement around 1900. The idea was that spirits were 4-D beings who could appear or disappear at any point, see everything, and so on. A fairly reputable astronomer, a Professor Zöllner, even wrote a book, *Transcendental Physics*, describing a series of seances he attended in an attempt to demonstrate that the "spirits" were actually 4-D beings. He seems, however, to have been hopelessly gullible, and his book is totally unconvincing. In general, the idea of a fourth dimension seems to precipitate authors into orgies of occultist mystification, rather than to lead to clear-sighted mathematical inquiry. The fact that something is difficult does not mean it has to be confused. The best of the books on the fourth dimension written from a mystical point of view is *Tertium Organum* by P. D. Ouspensky, who also has a good chapter on the fourth dimension in his book *A New Model of the Universe*.

In any case, Abbott's *Flatland* ends shortly after A. Square takes his "trip" into the third dimension. The Flatlanders lock him up and throw away the key. It has been this author's great good fortune to come into the possession of the true chronicle of the rest of A. Square's life.

A. Square had been in jail for about ten years when suddenly his old friend A. Sphere turned up again as a circle of variable size in poor Square's cell. "What's happening, baby?"

"Ah, noble Sphere, would that I had never seen you, would that I had been of too small an angularity to grasp your message."

"Man, you ain't seen nothing yet! You want me to lift you out of this jail and put you back in your wife's bedroom? Though I oughtta tell you, there's another mule kicking in your stall, a big sharp Isosceles."

"Sphere, Sphere, if only they'd believe me! There's no use letting me out. They'd just lock me up again, maybe even guillotine me. No, I knew you'd return and I have a plan. Turn me over. Turn me over and then my very body will be proof that the third dimension exists."

A. Square then explained his idea. He had been thinking about Lineland some more. Lineland was a world which Square had seen in a dream once, many years ago. Lineland consisted of a long line on which segments (Linelanders), with sense organs at either end, slid back and forth (Figure 24).

Fig. 24.

A. Square thought of Lineland in the same way in which we think of Flatland. He confronted his difficulties with the third dimension by imagining the Linelanders' difficulties with the second dimension. In jail for having preached the subversive doctrine of the third dimension, A. Square was understandably concerned with having A. Sphere create some permanent change in Flatland that would attest to the reality of the third dimension. (Note here that Prof. Zöllner was also concerned with getting the spirits to do something that would provide a lasting and incontrovertible proof of their four-dimensionality. His idea was a good one. He had two rings carved out of solid wood, so that a microscopic examination would confirm that they had never been cut open. The idea was that spirits, being free to move in the fourth dimension, could link the two rings without breaking or cutting either one. In order to ensure that the rings had not been carved out in a linked position, they were made of different kinds of wood, one alder, one oak. Zöllner took them to a seance and asked the spirits to link them, but unfortunately, they didn't).

In his cell A. Square had pondered on the kind of permanent change he could create in Lineland if he were back there. He could, of course, remove one of the segments, but this would probably just be termed a mysterious disappearance. He recalled that each Line-lander had a voice at each end, a bass on the left and a tenor on the right. If he turned one over, the voices would be reversed and everyone could observe this in Lineland (Figure 25). Now, if he could rotate a segment around a point, shouldn't A. Sphere be able to rotate a square around a line (Figure 26)? And everyone in Flatland would be able to tell, since everyone was built so that if the eye was toward the north, the mouth was toward the east. A. Sphere could turn A. Square into his own mirror image (Figure 27)!

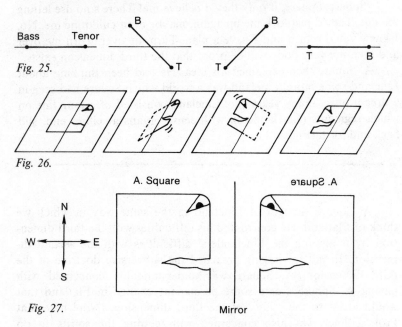

Fig. 25.

Fig. 26.

Fig. 27.

No sooner said than done. A. Square (or, rather, erauqS .A) called the guard: "Look, you dull-witted fool, I've rotated through the third dimension. I'm my own mirror image. Ha, ha, ha, ha! Show me to the High Priest! Now they will, they *must* believe!"

Well, the Flatlanders were quite impressed. They were so impressed they decided to put A. Square to death.

Before we continue this hair-raising tale, let's think about the analogy for us. It would seem that a 4-D being could turn us into our own mirror image by rotating us, in the fourth dimension, around a plane that cuts through our body—say the plane that includes the tip of your nose, your navel and your spine. What would it feel like to be rotated like this? How should I know? I can tell you a few things, though. One thing is the rather disgusting fact that when the rotation was only half completed, all of you that would remain in our 3-D space would be the plane around which the rotation was taking place. That is, you would look like a single vertical cross section of a human being—for if you look back at the pictures of A. Square's rotation, you can see that in the middle two pictures all that a Flatlander would see of him would be a cross section of his body. If the sphere had paused in the middle to pump A. Square up and down through the plane of Flatland, the guard would have been treated to a view of all the cross sections of A. Square's body. The same goes for you.

There is actually a model of how a rotation of a 3-D object through 4-D space might appear. Consider the picture of A. Cube, looking out at you from behind this sheet of paper, in Figure 28. His right eye is triangular, but his left eye is circular. Suppose that this sheet of paper were a mirror. In that case, A. Cube's mirror image would be on this side of the paper with its back to you (Figure 29). Note that there is no way in which you can move A. Cube in 3-D space so as to turn him into his mirror image, any more than you can turn yourself into your own mirror image by walking behind the mirror. Your heart will always be on your left, your mirror image's heart will always be on the right. But if we look at this picture (Figure 30), it seems to alternate between being A. Cube and A. Cube's mirror image. This figure, when drawn without the eyes (as in Figure 4), is called the Necker cube. If you look at a Necker cube for a while it spontaneously turns into its mirror image and back again. It you watch it "do" this often enough, the twinkling sort of motion from one state to the other begins to seem like a continuous motion. But this motion can only be continuous if it is a rotation in 4-D space. So perhaps we can actually produce a 4-D phenomenon in our minds! H. A. C. Dobbs has a paper in Fraser's book (see Bibliogra-

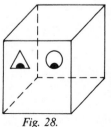

Fig. 28.

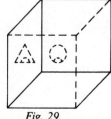

Fig. 29.

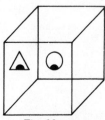

Fig. 30.

phy) in which he presents this argument and concludes that our consciousness is four-dimensional, with three space dimensions and one dimension of "imaginary time."

Let's get back to poor old A. Square. His pals were calling him an "object of horror to the gods," and they were getting ready to guillotine him. Our 3-D guillotine works by interposing a plane segment between two parts of the victim's body. The Flatland guillotine worked by interposing a line segment between two parts of the convicted polygon's body (Figure 31). Same difference.

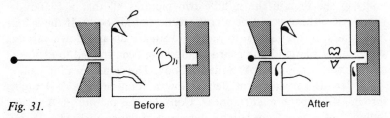

Fig. 31.   Before   After

A. Square was sweating it. He was so worried he hardly had time to enjoy being his own mirror image—writing backwards and stuff like that. He called out to the Sphere for help many, many times, but he only got static.

Finally, one grey, tight-stomached dawn, A. Square was led out to the "Splitting Field," where the guillotine was located. He saw many of his old friends, but none met his eye. His death sentence was read out, and two sharp Isosceles began dragging him toward the horrible instrument of destruction. And then, and then, and then ...

(1)

Complete this table:	Vertexes	Edges	Faces	Solids
Point	1	—	—	—
Segment	2	1	—	—
Square	4	4	1	—
Cube				
Hypercube				
Hyperhypercube				

(2) Locate the eight cubes that form the eight solids of the hypercube in Figure 5. It would be better to make several copies of this figure than to draw the cubes right in the book. A simple way of duplicating the drawing is to construct a regular octagon and then draw a square on the inside of each edge of the octagon.

(3) How many hypercubic feet of hypervolume are there in a hypercube each of whose edges is two feet long?

(4) The formula for the hypervolume of a hypersphere of radius r is obtained by evaluating the definite integral $\int_{-r}^{r} \frac{4}{3} \pi (\sqrt{r^2 - x^2})^3 \, dx$. Where does this integral come from? (Hint: Compare this integral to the integral $\int_{-r}^{r} \pi (\sqrt{r^2 - x^2})^2 \, dx$ which gives the volume of a sphere of radius r.)

(5) Suppose that every object in our space were an inch thick in the direction of the fourth dimension. Would we notice this 4-D component of our bodies' measurements? (Hint: Would A. Square notice if everything in Flatland were an inch thick in the direction of the third dimension?)

(6) Professor Zöllner attempted yet another experiment to demonstrate that "spirits" were free to move themselves and the objects of our space in a space of four dimensions. What Zöllner did was to place a snail shell on the table and ask the spirits to turn it into its mirror image. In what way does a spiral shell differ from its mirror image?

(7) Our actual retinal images of the world are 2-D. What sorts of visual experiences cause us to believe that our visible world is actually 3-D? How do you think A. Square manages to translate his 1-D retinal images into a mental image of a 2-D world?

(8) If you stuck the fingers of one hand through Flatland, A. Square would see you as five irregularly shaped objects, each covered with a tough pink hide. If a 4-D being stuck the "fingers" of one "hand" through our space, what would you see?

(9) The goal, in part, of the Cubist painters was to combine all the different possible views of an object into one picture. To what extent would a photograph of an object taken from a point not in our 3-D space accomplish this goal?

(10) Mystics have frequently maintained that our consciousness can be higher-dimensional. If we think of A. Square's normal 2-D thoughts as being network-like patterns on the 2-D space of Flatland inside his head, how can we represent his "higher-dimensional" thoughts? Why would it be difficult for him to communicate these thoughts to his fellow Flatlanders?

2

NON-EUCLIDEAN GEOMETRY

And then along came Sphere. As the executioner began to shove the "blade" of the guillotine home, A. Sphere fastened himself to a point between A. Square and the "blade" and began pulling upwards. He began stretching the space of Flatland, and he continued stretching it until the little space between A. Square and the "blade" had become big enough to hold the whole "blade."

The idea is that we imagine the space of Flatland to be a sort of elastic film that can be distorted by a pull in the direction of the third dimension.

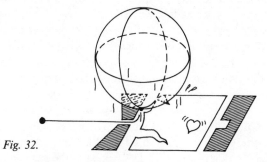

Fig. 32.

How is it that having the sphere pull up on a point between A. Square and the "blade" of the guillotine will make space more spacious? Consider the effect of pulling up at some particular point of Flatland's space (Figure 33).

We can see that if you take a point X between two points A and B, you can make the distance between A and B as great as you please by pulling up on X. In particular, by pulling up on a point in between the tip of the "blade" and A. Square, the sphere was able to make the

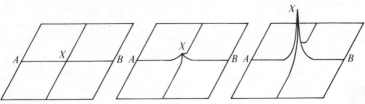

Fig. 33.

distance between A. Square and the "blade-guide" greater than the length of the "blade."

The Flatlanders were *very* impressed, and wouldn't you be if you saw a man survive being guillotined—survive because the blade couldn't manage to finish traversing the last inch to the man's neck? And after A. Sphere plucked out the heart of the High Priest, the Flatlanders went wild. "Free A. Square before we all get killed!" they cried, and so it came to pass that A. Square, once a condemned polygon, became a leading researcher at Flatland U.

A. Square was confused by his encounter with curved space. It had never occurred to him that space could be anything but flat. And, lest we sneer at this dimensionally impoverished creature's difficulties, we should wonder now if our 3-D space could be in any way "curved." We refer to a flat 2-D space as a plane, but so alien is the idea of curved 3-D space that we do not even have an English word to express the concept "noncurved 3-D space." Mathematicians sometimes call a noncurved 3-D space an E3 (analogously, they call a plane an E2 and a straight line an E1). The "E" is for Euclid, who first described the properties of flat space in a comprehensive way.

Let us all (you, me and A. Square) see what Euclid had to say about flat space. Euclid's system consists basically of five postulates and proofs of many propositions from these postulates. The five postulates consist of certain assumptions about the way points and straight lines behave in space. It is up to us to decide if these assumptions hold in the space in which we live. It is up to A. Square to decide if these assumptions hold in Flatland. As it turns out, asking if Euclid's postulates hold in a space is the same as asking if that space is "flat," or noncurved—whatever that might mean.

What *are* Euclid's postulates?

FIRST POSTULATE:
There is exactly one straight line connecting any two distinct points (see Figure 34). By "exactly one" we mean "at least one and no more than one." What do we mean by "straight line"? Realistically, we have a pretty good idea of what a straight line is in our space: the shortest path between two points. However, in order to start with as

little as possible in the way of assumptions, we do not make any initial assumptions about what straight lines are. The only properties of straight lines that we will assume will be those provable from the postulates we decide to accept.

SECOND POSTULATE:

Every straight line can be continued endlessly (see Figure 35). This has to do with our feeling that space has no boundaries, no edges. You never reach a point beyond which a given straight line cannot be continued.

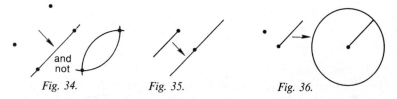

Fig. 34. Fig. 35. Fig. 36.

THIRD POSTULATE:

It is possible to draw a circle with any given center and radius (see Figure 36). This postulate does not, on the face of it, seem to be about points and straight lines. And what is "circle" supposed to mean here? We cannot do better than Euclid's definition: "A circle is a plane figure contained by one line such that all the straight lines falling upon it from one point among those lying within the figure are equal to one another. And the point is called the center of the circle." But what does the ability to draw circles have to do with the properties of space? We might be inclined to think that the ability to draw circles depends on owning a compass, rather than on some fundamental property of space with respect to its straight lines and points! But does not the fact that a compass works properly have something to do with space? How do you *know* that a compass draws a circle—that is, how do you *know* that the separation between the stabbing and drawing ends of the compass remains the same as you rotate it about the stabbing end? The idea seems to be that a material body (or an imagined line segment) does not change its size as we move it about in space. Thus, part of what the Third Postulate says is that distance in space is to be defined in such a way that a line segment's length does not change when we move it from one place to another.

FOURTH POSTULATE:

All right angles are equal to each other (Figure 37). The content of this postulate is not clear until we have defined right angles: "When a straight line set up on a straight line makes the adjacent angles equal to one another, each of the equal angles is right." This postulate seems to be equivalent to the assumption that the things we are calling "straight lines" don't have any corners. Another way of expressing this is that the Fourth Postulate says that space is "locally flat," that is, that a small enough region of space will not betray any curvature.

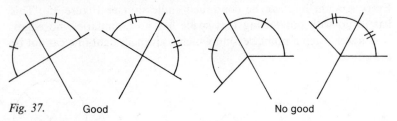

Fig. 37.   Good   No good

FIFTH POSTULATE:
Given a line *m* and a point *P* not on *m*, there is exactly one line *n* that passes through *P* and is parallel to *m* (Figure 38). It is understood here that lines are said to be parallel when they do not intersect. The Fifth Postulate could fail to hold in two different ways. It might be that there were no lines through *P* parallel to *m*, or it might be that there was more than one line through *P* parallel to *m*. It turns out that both alternatives are possible, if we choose the right kind of "straight lines."

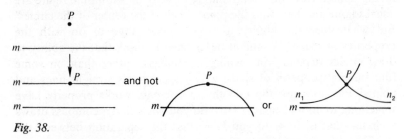

Fig. 38.

In general, we accept Euclid's first four postulates for our space. It is certainly true that given two nearby points, there is just one shortest path from one to the other. It is certainly true that there seem to be no boundaries to space. It is certainly true that objects do not seem to expand or contract as we move them around. And it is certainly true that our "straight lines" do not have corners in them. But the Fifth Postulate is not so easy to accept on the basis of experience. Might it not be that even lines that start out looking parallel come together slowly as they get farther and farther away from us? Or, conversely, might it not be possible that lines that start out looking as if they will intersect bend away from each other slowly as they are produced out toward infinity, perhaps approaching each other asymptotically, but never actually intersecting?

For many centuries people believed that it was not possible for the Fifth Postulate to be false in our space. There were two sorts of reasons given for this belief. The first reason was that God would not have botched his work. The idea was that space was an almost divine, eternally existing absolute form. As such, it would certainly not be expected to contain vilely converging and diverging collections of straight lines of the type required to violate the Fifth Postulate. The second reason given for the flatness of space is essentially due to Immanuel Kant, the German philosopher. Kant wrote at a time when the kind of authoritarian theological standpoint embodied in the first argument for the Fifth Postulate was losing ground. His argument for the truth of the Fifth Postulate was that space is largely a creation of our own minds, that we cannot imagine non-Euclidean space, and hence space is Euclidean (i.e., satisfies the Fifth Postulate). The argument that space is a creation of our own minds is an interesting one; the idea is that we cannot see or imagine seeing anything that is not located in space. Space, to use Kant's phrase, is "an ineluctable modality of our perception." Space may not have any "real" existence, but there is no way in which we can order our sense perceptions without using the organizing framework of space. This is fine, but why shouldn't we be able to imagine non-Euclidean space? Kant thought we couldn't because no one *had* at the time when he wrote (around 1780). So he concluded that our space must necessarily satisfy the Fifth Postulate, since the alternatives were unimaginable.

Kant, however, was wrong. We *can* imagine non-Euclidean spaces. Let's start with a space where there are no parallel lines. Rather than working with 3-D space, let's make it easy for ourselves and start out with 2-D space; that is, let's describe a version of Flatland where every two lines meet somewhere.

The idea is to let Flatland constitute the surface of a large sphere (Figure 39). A. Square and his cohorts are curved so as to stick to the surface of the sphere. They can slide around on it to their hearts' content, and we can imagine that they have not even noticed yet that their space is anything other than the infinitely extended plane which they imagine it to be. Postponing their momentous discovery that something is amiss, let us see if the Fifth Postulate holds in a spherical (rather than flat) 2-D space.

No, it does not. Why not? Well, first of all, what exactly is meant by "straight line" on the surface of a sphere? Obviously, any line that is contained in the surface of a sphere cannot be "really"

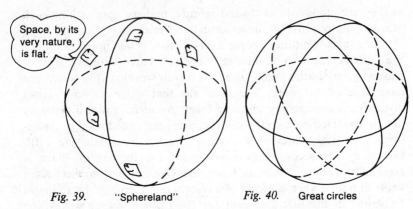

Fig. 39.   "Sphereland"   *Fig. 40.*   Great circles

straight, and we are not allowed to tunnel under the surface to produce lines such as the diameter of the sphere (since such lines do not lie on the *surface* of the sphere; as far as A. Square is concerned, the space inside and outside the sphere's surface *does not exist*). Now, which of the lines that you can draw on a sphere's surface is the straightest? That is, if A. Square and his friend Dr. Livingchip took a thread and pulled it taut, what kind of line would the thread lie on? Look at a globe map of the world. On this globe the longitude lines and the equator look straight, but the latitude lines look curved. There is no way you can draw a line straighter than the equator on the sphere.

The lines on the sphere that we call "straight" are the so-called great circles: "great" because there's no way you can make a great circle bigger (Figure 40). Slide the equator north *or* south and it's going to shrink. A great circle on a sphere has the same radius as the sphere, and you can't beat that. If A. Square travels along a great circle he doesn't feel as if he's curving off to the left or to the right. He *is* curving, but only in the direction of the *third* dimension, that is, in a direction perpendicular to his two space dimensions.

Now, the point of all this was to get a space where the Fifth Postulate fails. It fails on the sphere when you take great circles to be "straight lines," since every two great circles intersect each other. Say you take a great circle *m* and a point *P* not on *m* and try to find a great circle that goes through *P* and never hits *m* (that is, is parallel to *m*). No can do!

Another unusual aspect of geometry on a sphere is that the First Postulate does not hold there, either. If, for instance, you take the north and the south poles, there are infinitely many great circles connecting these two points, and not just one (Figure 41). (It is possible to have no parallel lines without giving up the First Postulate —see Problem 3.)

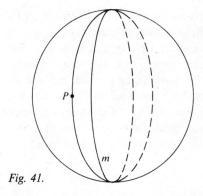

Fig. 41.

Notice that we obtained our model of the "no-parallels postulate" by taking a curved space and letting our "straight lines" be what are called the *geodesics* of the surface. A geodesic line on a curved surface is a line that is as straight as possible, the kind of line you get if you pull a thread (which cannot snap out of the surface) taut on the surface. The great circles are geodesics on the sphere. What if, instead of taking a curved space and straight "straight lines," we take a flat space and *curved* "straight lines"?

In other words, what we'd like to do now is to find a collection of curved lines in the regular plane such that, if we start pretending these lines are "straight," then we'll get something that behaves just like the sphere with its great circles. No problem. Here's what you do.

Take the plane and add a point at infinity. The idea is that if you go out in any direction at all forever, you end up at the point at infinity. Now draw a nice big circle on your plane and call this the Fundamental Circle. You're going to claim that the Fundamental Circle is a "straight line." What else is going to be a "straight line"? First of all, any straight line that goes through the center of the Fundamental Circle. Note that any two of these straight lines meet in two points, the center of the Fundamental Circle and the point at

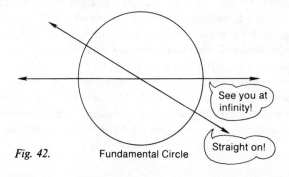

Fig. 42.   Fundamental Circle

infinity. Since we only have one point at infinity, all four arrows here meet at this one point at infinity (Figure 42).

Secondly, we'll call any circle that intersects the Fundamental Circle in two diametrically opposite points a "straight line." As well as studying Figure 43, it would be a good idea for you to get out a compass, draw yourself a Fundamental Circle and draw a number of circles that cut your Fundamental Circle in two diametrically opposite points. Given that these circles are "straight lines," what kind of space do we have here? Just to have a name for it, let's call it a Flat Sphere.

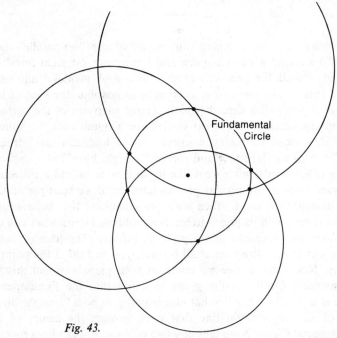

Fundamental
Circle

Fig. 43.

Notice that, given two diametrically opposite points on the Fundamental Circle, there are lots of "straight lines" through these two points. We have labeled these lines according to the position of their centers on a *y*-axis which we can imagine as having its zero-point at the center of the Fundamental Circle (Figure 44).

Observe that all the "straight lines" in Figure 44 are the kind of lines you would expect to be parallel if you just looked near the *y*-axis, but that they all meet each other. In other words, the plane (plus the point at infinity) with these "straight lines" is another model of non-Euclidean geometry. Just as on the sphere, the "no-parallels postulate" holds and the First Postulate fails.

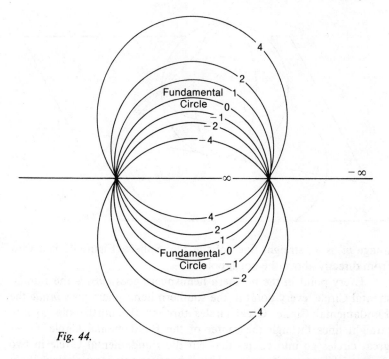

Fig. 44.

What is the relation between the space called the Flat Sphere and the surface of a real sphere? They are *isomorphic*. That is, we can find a one-to-one mapping from the set of points on the sphere onto the set of points on the plane (plus the point at infinity) such that every "straight line" on the sphere is taken into a "straight line" of the Flat Sphere. What is the mapping? *Stereographic projection*. It works like this.

Take a sphere and set it down on a plane. Set it down so that the south pole of the sphere is the point where the sphere touches the plane. Now, given any point *P* on the surface of the sphere, draw a straight line *NP* from the north pole to the point *P* and continue this line until it cuts the plane. Call the point where the continuation of *NP* cuts the plane *P'*. *P'* is the image of *P* under stereographic projection (Figure 45).

Notice that every point *P* on the sphere has a unique image *P'*. The image of the south pole is the point where the sphere touches the plane. The image of the north pole is the point at infinity. We can find the images of the great circles on the sphere by letting the image of a great circle *m* be *m'*, where *m'* is the set of all points *P'* such that *P* lies on *m*. Let the image of the equator be the Fundamental Circle on the plane. Observe that, given any great circle *m* on the sphere, its

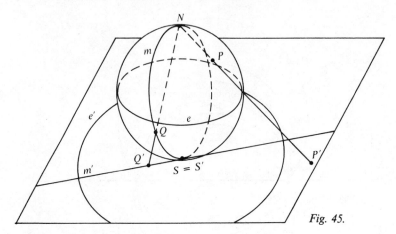

Fig. 45.

image *m'* is a "straight line" of the Flat Sphere. Figure 46 is a view from directly above the north pole.

Every point in the northern hemisphere goes *outside* the Fundamental Circle, every point in the southern hemisphere goes *inside* the Fundamental Circle. Great circles through the north pole go into straight lines through the center of the Fundamental Circle. Other great circles go into circles that cut the Fundamental Circle in two diametrically opposite points. This is because any great circle will cut the equator in two diametrically opposite points. The images of these points are diametrically opposite points on the Fundamental Circle (which is the image of the equator).

What we have indicated here is that the real sphere and the Flat Sphere are isomorphic spaces. When we speak of a "space" we mean a collection of points and a collection of "straight lines." When we refer to the sphere as a space, we are thinking of the collection of points on the sphere as our points and the collection of great circles on the sphere as our "straight lines." When we refer to the Flat Sphere as a space, we are thinking of the collection of the points on the plane (plus the point at infinity) as our collection of points and the collection of lines and circles that cut the Fundamental Circle in two diametrically opposite points as our collection of "straight lines." When we have two isomorphic spaces, such as the sphere and the Flat Sphere, we can conclude that there is no way in which an inhabitant of the one or the other can decide which one he is "really" in. That is, A. Square might be able to learn that his space was "spherical," but there would be no way in which he could determine if his space was a real sphere or a Flat Sphere.

What *is* the difference between the real sphere and the Flat

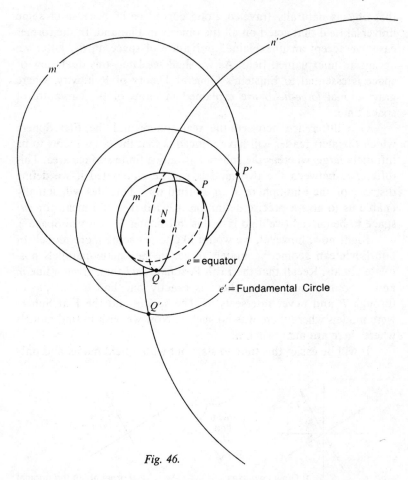

Fig. 46.

Sphere? The sphere is a *curved* surface whose "straight lines" are geodesics (that is, lines that are as short as possible; in general, on any surface one can find the geodesic between P and Q by stretching a thread from P to Q so that the thread lies entirely on the surface and is as taut as possible). The Flat Sphere is a *flat* surface whose "straight lines" are curved. In the one case we have curved space and straight lines, in the other case we have flat space and curved lines. The first type of model is called a *curved-space model*, the second is called a *field model*. That is, we would think of Flatland as being the real sphere if we argued that (for some reason) the space of Flatland was *curved* and that objects naturally traveled along geodesic (shortest) paths. On the other hand, we would think of Flatland as being the Flat Sphere if we argued that the space of Flatland was *flat*, but that objects naturally traveled along curved paths because of some universal field that acted on all the objects in Flatland. In the former case, we accept an unexplained curvature of space, in the latter we accept an unexplained field. As we shall see later, this dual view of space is essential to Einstein's General Theory of Relativity, where gravitational *force-fields* are explained in terms of the curvature of space-time.

One difference between the real sphere and the Flat Sphere which the alert reader will have noticed is that the latter seems to be infinitely large, whereas the sphere has only a finite surface area. This difference between the two models can be eliminated if we define distance on the Flat Sphere in an unusual way. This idea will actually enable us to give a precise definition of what it would mean for *our* space to be curved (we'll go into this in Chapter 3, "Curved Space").

Right now, however, we would like to get some more models of non-Euclidean geometry, models in which the "many-parallels postulate" holds. Recall that the Fifth Postulate said that, given a line *m* and a point *P* not on *m*, there is exactly one line *n* that passes through *P* and never intersects *m*. The sphere and the Flat Sphere were models where there was no such *n*. Now we wish to find models where there are many such *n*.

It will be easier this time to start out with a field model and only

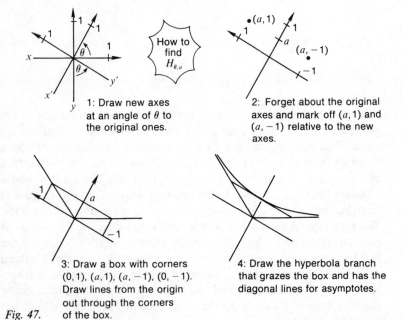

How to find $H_{\theta,a}$

1: Draw new axes at an angle of θ to the original ones.

2: Forget about the original axes and mark off $(a,1)$ and $(a,-1)$ relative to the new axes.

3: Draw a box with corners $(0,1)$, $(a,1)$, $(a,-1)$, $(0,-1)$. Draw lines from the origin out through the corners of the box.

4: Draw the hyperbola branch that grazes the box and has the diagonal lines for asymptotes.

Fig. 47.

then attempt to find the related curved-space model. We will call our space the Flat Saddle. The points of the Flat Saddle are all the points on the plane, and the "straight lines" of the Flat Saddle are hyperbolas of a certain special type.

For every angle θ such that $0° \leqslant \theta < 360°$ and every real number a such that $0 \leqslant a$, we let $H_{\theta, a}$ be the line that is the rotation of the right-hand branch of the hyperbola $(x^2/a^2) - y^2 = 1$ through θ degrees counterclockwise. Thus, to draw $H_{\theta, a}$ we first draw new x and y-axes at an angle of θ to the old ones, then we draw asymptotes through the origin and the points $(a, 1)$ and $(a, -1)$, and then we draw the hyperbola with these asymptotes which passes through $(a,0)$ (see Figure 47).

Note that if $a = 0$, then the hyperbola $(x^2/a^2) - y^2 = 1$ is just the y-axis. Thus our "straight lines" will be real straight lines that pass through the origin as well as certain types of hyperbolas. Why didn't

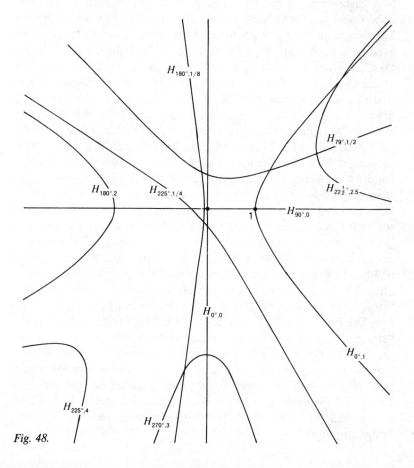

Fig. 48.

we just take *every* hyperbola to be a "straight line"? Because then there would be many "straight lines" between any two points, thus violating the First Postulate. As it turns out, I have been able to prove, with the help of Professor Paul Schaefer, that, given any two points in the plane, there is exactly one $H_{\theta,a}$ that passes through the two given points. Thus the Flat Saddle is a model of the First Postulate (Figure 48).

The Flat Saddle is also a model of the Second Postulate since each of its "straight lines" does go on forever in both directions. The Flat Saddle is *not* a model of the Third Postulate in the sense that a compass draws a curve all of whose radii are equal—since "straight lines" going out from a point in various directions are curved in various ways. The Third Postulate *does* hold in the sense that, given any direction, we should be able to measure off to a point at a distance of r along a "straight line" in that direction. The question of how distance is to be measured in field spaces such as the Flat Sphere and the Flat Saddle is a touchy one. The problem is that we have a natural feeling that *"straight lines" should be geodesics*. It could even be argued that *this* is the content of the Third Postulate, if we take the Third Postulate to say: "If you take a length of string and attach one end of it to a point P and swing the string around, holding it stretched as tight as possible, then the free end of the string will draw a curve c which is a circle—a circle in the sense that if we take any two 'straight lines' that pass through P, then the segments of the 'straight lines' that lie between P and c are all equal"(Figure 49). This version of the Third Postulate fails for the Flat Sphere and the Flat Saddle if we assume that distance in these spaces is measured in the same way as on the plane. There is nothing, however, to prevent us from defining distance in *different* ways in these spaces, as we shall see in the next chapter. If distance is defined in a suitable way, our "straight lines" *will* be geodesics! A stretched string will be along a geodesic line that is also a "straight line." Swinging a string around will produce a closed curve that satisfies the definition of a circle, although this curve will not resemble a circle any more than the "straight lines" resemble straight lines (Figure 50).

The Fourth Postulate will hold on the Flat Saddle since all of the "straight lines" are smooth (differentiable) curves. A way in which the Fourth Postulate could fail would be for us to work with a curved-space model that had a little peak on it somewhere. At such a peak (like the one A. Sphere made when he pulled on Flatland) two lines can cross each other and make four equal angles, each of which is less than 90°! The fourth postulate says that space has no such "singular points."

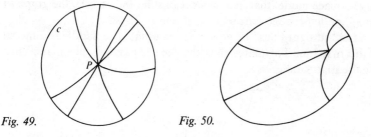

Fig. 49.   Fig. 50.

So Euclid's first four postulates seem to hold in the space called the Flat Saddle. How about the Fifth? This fails because, given a "straight line" *m* and a point *P* not on *m*, we can find many hyperbolas of the correct form that pass through *P* and miss *m* (Figure 51).

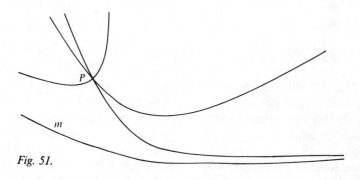

Fig. 51.

We arrived at the Flat Sphere by starting out with the real sphere and then thinking of stereographic projection (the idea for the Flat Sphere is from Hans Reichenbach's book, *The Philosophy of Space and Time*). But actually, we arrived at the Flat Saddle *without* first thinking of some curved-space model. The idea behind the Flat Saddle is that you imagine yourself to be standing at the origin of a weird space. Your lines of sight—that is, lines that go through the origin—are straight, but lines that do not go through the origin appear to bend away as they go out toward infinity. This effect is more pronounced for lines that are further away from you. If you were standing in a corridor in this space, you would see yourself as standing in the neck of a horizontal hourglass (assuming that your perceptions had not yet adapted to this new space).

Is there a curved-space model that is related to the Flat Saddle in a way similar to the way in which the real sphere was related to the Flat Sphere? To be quite honest, I am not sure. Let me describe a curved-space model that *may* work. Consider the Saddle, the graph of $z = xy$, a "hyperbolic paraboloid" (Figure 52).

Let "straight lines" be geodesics, as usual. The Saddle *is* a model of the first four Postulates plus the "many-parallels postulate," just like the Flat Saddle.

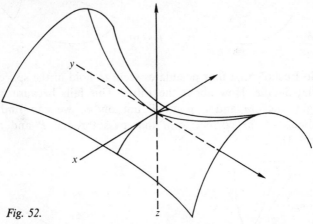

Fig. 52.

The difficult question is whether the Saddle is isomorphic to the Flat Saddle. The obvious map would be to simply let each point (x, y, z) that lies on the Saddle go into the point (x, y) on the Flat Saddle. In other words, we just project each point of the Saddle straight down (or up) onto the xy-plane. Does this map take the geodesics of the Saddle into the "straight lines" of the Flat Saddle? If not, can the map be fixed up to work right? These are difficult questions, but it seems safe to say that if there is any curved-space model at all that is isomorphic to the Flat Saddle, then it will look something like the Saddle.

How would A. Square know what kind of space he lived in? It would be hard, perhaps, for him to test the Fifth Postulate. On the Flat Saddle, for instance, he could go and go and go, watching two lines that looked as if they *ought* to intersect eventually, but he would never know if they really did fail to intersect, or if he just hadn't gone far enough out. What we are getting at is this question: Are there any *local* properties of space that determine which of the three "parallel" postulates the space satisfies?

The answer is yes. Using the Fifth Postulate, it is possible to prove that the sum of the angles in any triangle is 180°. Using the Fifth Postulate, it is possible to prove the Pythagorean Theorem. It turns out that both these proofs fail when the Fifth Postulate is false. We'll tabulate all the relevant information in a minute, but first let me introduce a basic distinction between surfaces such as the sphere and surfaces such as the Saddle. A surface is said to have *positive curvature* if it is concavo-concave or convexo-convex; a surface is said to have *negative curvature* if it is concavo-convex or convexo-concave. What does this mean? Take a surface and pick a point on it. Draw two lines on the surface that cross each other at right angles at this point and such that at least one of the lines is as curved as possible. If the two lines are curved in the *same* direction (both upward or both downward), we say the surface has positive curvature at this point. If the two lines are curved in opposite directions (one upward and one downward) we say the surface has negative curvature at this point. The sphere has positive curvature at each of its points, the Infinite Peak has negative curvature at each of its points. The surface $z = x^2 + y^2$ has positive curvature at each of its points, the $z = \dfrac{1}{x^2 + y^2}$ has negative curvature at most of its points. What does it mean when a surface has *zero curvature* at a point? This means that at least one of the two lines mentioned above is *really* straight. A cylinder, for example, has zero curvature at each of its points. (See Figure 53 for illustrations.)

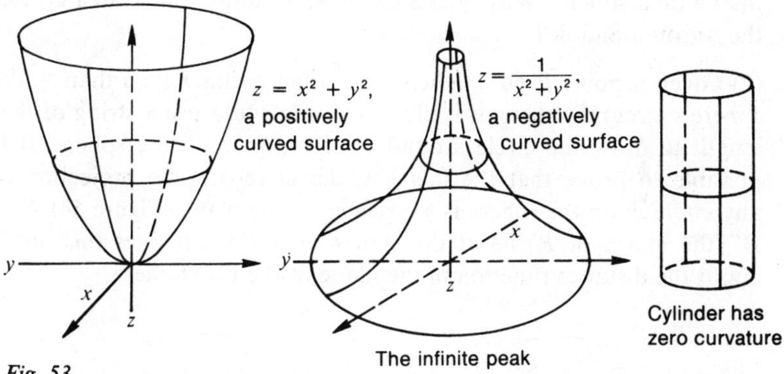

$z = x^2 + y^2$,
a positively
curved surface

$z = \dfrac{1}{x^2 + y^2}$,
a negatively
curved surface

Cylinder has
zero curvature

The infinite peak

Fig. 53.

On the next page is a table presenting the relation between the type of space one is in and the various properties which that space can have.

A curved-space model of this type has ____ curvature.	Given a line *m* and a point *P* not on *m*, there are ____ lines through *P* parallel to *m*.	The angle sum of every triangle is _____.	The square of the hypotenuse of a right triangle with sides *a* and *b* is _____.	The circumference of a circle with diameter 1 is _____.
Positive	No	> 180°	< $a^2 + b^2$	< π
Zero	One	= 180°	= $a^2 + b^2$	= π
Negative	Many	< 180°	> $a^2 + b^2$	> π

PROBLEMS ON CHAPTER 2

(1) Why does a right triangle with sides 3 and 4 have hypotenuse less than 5 on a sphere? Why does a circle with radius 2 have area greater the 4π on a Saddle?

(2) Given a point *P* on a sphere and some radius *r* (less than $\frac{1}{4}$ the sphere's circumference), it is clear that one could use a string of that length to draw the circle around *P* of radius *r* on the sphere. It is possible to prove that the image under stereographic projection of any circle *m* on the sphere is a circle *m'* in the plane (Figure 54). Will *P'* (the image of *P*) lie at the center of *m'*? What does this imply about the distance function in the plane of the Flat Sphere?

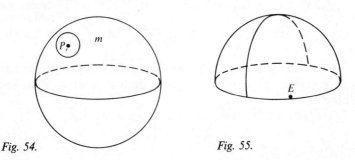

Fig. 54.   Fig. 55.

(3) I mentioned that it is possible to have no parallels and still have the first Postulate hold. One way of doing this is to take a *hemisphere*

and identify certain pairs of points on the edge. That is, for each point E on the equator (Figure 55) we choose another point E^* on the equator and pretend that E and E^* are the same point. What rule for associating E and E^* should one use?

(4) Look in a mirror and imagine how you would look if you painted all the negatively curved portions of your face blue and all the positively curved portions red.

(5) Draw a Fundamental Circle, pick two points P and Q, and carry out the construction of the "straight line" between two points on the Flat Sphere, as shown in Figure 56.

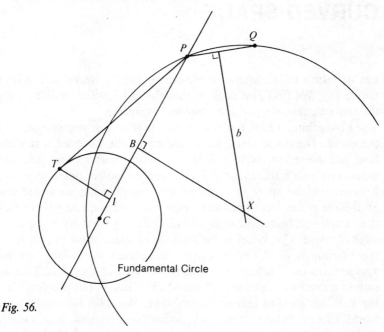

Fig. 56.

How to find the "straight line" between two points on the Flat Sphere.
(1) Given: Fundamental Circle and P and Q.
(2) Draw the segment PQ and construct this segment's perpendicular bisector b.
(3) Draw line PC.
(4) Construct the tangent PT to the Fundamental Circle.
(5) Construct the perpendicular TI to PC.
(6) Find the point B such that $CB = \frac{1}{2} IP$.
(7) Construct a line BX perpendicular to CP.
(8) The circle with center X and radius XP is the "straight line" between P and Q.

3

CURVED SPACE

Let us return to A. Square, Professor of Higher Space at Flatland University. We find him sunk in thought in his office, mulling over the fantastic discovery of the explorer Livingchip.

Livingchip had set out earlier that year to discover the edge of the world. The Flatlanders, we should point out, believed that Flatland was the region inside a circle of a radius of about one thousand miles, one year's travel for a Flatlander. In earlier times they had believed that the space of Flatland was infinite, what we would call an infinite plane, but in modern times they had come to believe that their space was finite, although no one could say what lay beyond the edge of space. This belief in the finitude of space came in part from the utterances of A. Sphere, who had taken to making regular appearances on Saturday nights at the Church of the Third Dimension in downtown Flatsburg. "Your world is round, flat peoples, it's a big ball like me and almost as han'some. We solid folks calls your world Etheric Sphere #666." A. Sphere's utterances were usually cryptic, but this one seemed clear enough: the space of Flatland was the inside of a large circle.

Something about this reasoning seemed wrong to A. Square, but these things were so hard to think about. And, after all, it would be impossible for the space of Flatland to actually be a sphere like A. Sphere ... or so it seemed until Livingchip returned from his journey to the "edge of space." Livingchip had set out due east two years earlier. The idea was that after about a year's travel he would reach the edge of the world. Once there he'd find out what it looked like, take some pictures, conduct some experiments, maybe chip a piece of it off, leave a Flatflag with the High Priest's name on it and come home.

Livingchip returned after two years all right, but he returned from the west instead of the east. This would not have been so surprising if Livingchip had not insisted that he moved in a straight line for two whole years without ever coming to the edge of space, that he had never turned back and never deviated from a straight path. The High Priest suggested that Livingchip be put to death, but the office of High Priest had become largely ceremonial since the day of A. Square's escape. The public wished to understand Livingchip's feat, not obliterate it, and they turned to A. Square for an explanation.

You, the reader, should not be surprised to hear that he came upon the idea that the space of Flatland was the surface of a sphere. But rather than running through the whole song and dance in 2-D terms, let's up it a dimension and see what it would be like if our 3-D space was the hypersurface of a hypersphere.

First of all, we would be able to duplicate Livingchip's feat. If we took off from the North Pole in a rocketship and continued flying straight away from the Earth long enough, then we'd see a nice-looking planet ahead of us after a while—and when we landed we'd find ourselves at the South Pole.

Note that it is just as easy to imagine a "spherical" Lineland as it is to imagine a spherical Flatland (Figure 57). Why is it so hard to imagine *our* space as "spherical"? The reason is that the curvature of our 3-D space would be in the direction of the fourth dimension. Our "straight lines" would actually be curved, but in a direction unknown to us. This becomes clearer if we consider a great circle on a sphere, say the equator. If A. Square slides along the equator he will say, "This line is straight; it bends neither to the left nor to the right. If it is truly curved it can only be curved in the direction of the mysterious third dimension." Similarly, a line in our space may appear to bend neither left nor right, neither toward us nor away from us, but may still be bent in the direction of the fourth dimension.

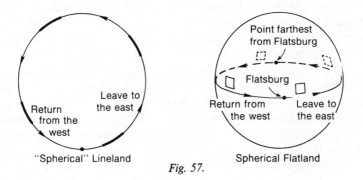

Fig. 57.

If our space was hyperspherical, we could actually detect this without flying around the universe, for, as we learned in the last chapter, any triangle that we drew with straight sides would actually have more than 180° in it. Unless, however, the radius of the hypersphere whose hypersurface forms our 3-D space were very small, this type of deviation would be too small to be noticed.

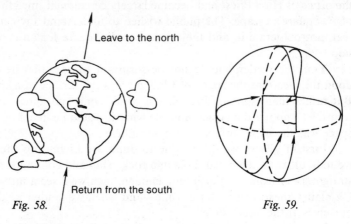

Leave to the north

Return from the south

Fig. 58.

Fig. 59.

It is interesting to imagine what it would be like to occupy the hypersurface of a rather small hypersphere, one with a circumference of, say, 50 yards. If you were floating in such a 3-D space, then no matter which direction you traveled in, you would return to your starting point after 50 yards. Imagine yourself to be floating in such a space. There is no matter besides you and some air, and you are equipped with a handheld jet to propel you. To start with, you are in a position very similar to that of an astronaut hanging in outer space. The difference is that if you jet away from your starting point in a straight line, you come back to your starting point after 50 yards. What do you see? It would seem that any direction you looked in, you would see yourself. Why? Well, what would A. Square see if he lived on a fairly small sphere? Whichever way he looked, he'd see himself (Figure 59). He would see a very large A. Square at a distance of about 50 yards from himself. The image he sees is actually even stranger than that, as we will see in a few pages.

But, returning to you floating in your tidy little spherical space and seeing a huge image of yourself at a distance of 50 yards, let us supply you with a large and stretchable balloon and conduct a new experiment. Imagine that you crawl inside this deflated balloon and begin to inflate it. You do this, let us say, by releasing compressed air from a tank of compressed air that you happen to have with you. The balloon begins to expand and you find yourself at the center of an expanding rubber sphere. A strange thing happens, though, when the sphere's diameter gets to be 25 feet. The rubber wall that separates you from the space outside the balloon stops being curved toward you and begins to appear flat. You are somehow enclosed by a wall that is completely flat, without curves or corners! When you release more compressed air from your tank, the wall begins to curve *away* from you, and soon the balloon which you started out inside of has become a balloon which you are outside of. The picture of Donald Duck which was on the outside surface of the balloon is now on the inside surface of the balloon. The balloon has apparently turned inside out, without being torn or punctured. You have passed from the inside to the outside of the balloon without going through its wall. Puzzled and slightly freaked-out, you turn to A. Square.

Say that the 2-D space of Flatland comprises the surface of a sphere with a circumference of 50 yards. Your crawling into a balloon corresponds to A. Square's getting inside some elastic closed curve. The moment when the 2-D balloon has expanded to become a great circle corresponds to the moment when your balloon appeared flat (Figure 60).

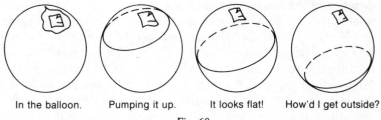

| In the balloon. | Pumping it up. | It looks flat! | How'd I get outside? |

Fig. 60.

The idea that our 3-D space may be spherical is not science-fiction, but rather an idea that is seriously believed by many responsible scientists. Albert Einstein was one of the first people to put this idea forward. What is the appeal of this idea? It is perhaps that it enables us to have a space that is not infinite but that is also without boundaries. We certainly would not want to have boundaries to our space. The very idea hardly makes sense, for if you could get to a point on the boundary, what would stop you from going further? On the other hand, there is something in us that recoils from the idea of a space that *goes on forever*, populated with infinitely many stars, infinitely many planets, infinitely many civilizations. If our 3-D space makes up the hypersurface of a hypersphere, however, we can have unbounded, but finite space. But wouldn't the point at the opposite end of the universe be a sort of boundary? Not really; if you were at this point you would be perfectly free to move in any three-space direction you could think of. It would just be that every one of those directions was toward Earth. (Similarly, if you are in Australia, you're free to sail a ship in any two-space direction you like. It's just that every one of those directions is toward the U.S.A.).

A natural idea at this point is that just as there could be many spherical Flatlands floating in 3-D space (Figure 61), there could be many hyperspherical Universes floating in 4-D space. Why can't we get off our hypersphere?

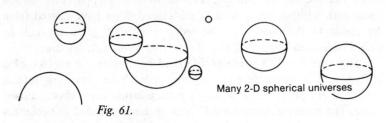

Many 2-D spherical universes

Fig. 61.

The problem is that in order to move in the direction of the fourth dimension we would have to be able to exert a force in the direction of the fourth dimension, and this we cannot do. No matter what A. Square does, he's only going to slide around on his sphere.

While we're on the subject of distinct 3-D universes floating in 4-D space, let us mention the idea of "parallel universes" that one occasionally reads about in science-fiction and in occult writings. Forget about the idea of curved space for now and just go back to the idea of Flatland as an infinite plane. The "parallel-universe" concept is that there would be two, or seven, or infinitely many Flatlands parallel to each other. In some versions people move from one parallel universe to another until they find one that suits them, the idea here being that every possible state of affairs is realized in at least one of the many parallel universes. In other versions we exist simultaneously in each of the universes; the "astral plane," for instance, is viewed as a parallel universe in which our "astral bodies" live (Figure 62). The astral body sometimes just copies the actions of our physical body, but sometimes—as when we dream—our astral body acts independently of the physical body. Guys who work on it are said to be able to "wake up" while they are sleeping and actually do things on the astral plane, such as travel to distant places and bring back reports of what is going on there. To what extent your astral body is connected to your physical body is unclear in the writings I have consulted. There was a great deal of interest in these ideas in the early part of this century, and recently there has been somewhat of a revival in occultist studies. Most of what I have read seems, however, to contain a large amount of wishful thinking. As life becomes less adventurous in our industrialized society, many people try to find new paths into the unknown. Perhaps we *are* actually 4-D beings and our physical bodies are only a 3-D cross section of our full bodies, but it cannot be said that there is any convincing evidence of this. Convincing evidence would consist of some consistent and plausible extension of our present theory of physics that would assume the four-dimensionality of ordinary physical bodies and predict verifiable experimental results. As long as there is no good *theory* of astral bodies, psychic phenomena and so on, no experiment can be really convincing.

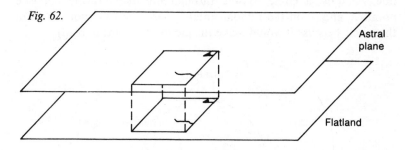

Fig. 62.

Astral plane

Flatland

Going back to the idea of hyperspherical universes floating in 4-D space, notice that we could move onto higher levels by asserting that the 4-D space in which our hypersphere floated was actually curved into the hyperhypersurface of a hyperhypersphere floating in 5-D space, that there were many such hyperhyperspheres, that the 5-D space was actually the hyperhyperhypersurface of a hyperhyperhypersphere floating in 6-D space, and so on and on. Once we start adding dimensions there is no logical stopping place short of infinity. Should the infinite-dimensional space be curved or flat? Mathematicians have, in a different context, actually studied a sort of flat infinite-dimensional space called Hilbert space. But, to quote Whately Smith, "The nature of maximally dimensional space is a question which I do not propose to discuss here as it is somewhat conspicuously outside the sphere of practical politics."

Just as it was possible to get a Flat Sphere isomorphic to the real sphere by taking your points to be the points on the plane plus the point at infinity, and your "straight lines" to be circles and lines of a certain type, it is possible to get a Flat Hypersphere by taking your points to be the points in regular 3-D space, your "planes" to be spheres and planes of a certain type, and your "straight lines" to be

circles and lines of a certain type. Here's the way it works. Choose a
nice-looking sphere in your regular 3-D space and call it your
Fundamental Sphere. Now say that a "plane" is (i) any plane that
passes through the center of the Fundamental Sphere and (ii) any
sphere whose intersection with the Fundamental Sphere is a great
circle of the Fundamental Sphere. A "straight line" is (i) any line that
passes through the center of the Fundamental Sphere and (ii) any
circle that cuts the Fundamental Sphere in two diametrically opposite
points. This model is described in Hans Reichenbach's excellent
book, *The Philosophy of Space and Time.* Notice that any two
"planes" intersect each other in a "straight line." Thus, in Figure 63,
"planes" P and Q are spheres whose intersection will be a "straight
line," that is, a circle passing through the diametrically opposite
points X and Y on the Fundamental Sphere (we have not drawn in
this circle because it would make the picture too hard to read).

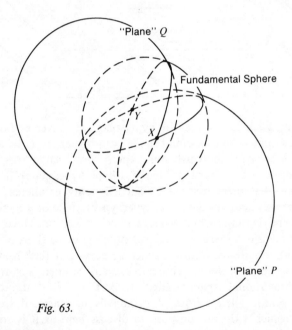

Fig. 63.

We can imagine the Flat Hypersphere as being isomorphic to a
real hypersphere under a 4-D stereographic projection. You would
take a hypersphere and an E3 that had one point in common, called
S. Let N be the point on the hypersurface of the hypersphere that
was as far away from S as possible. Given any point P in the 3-D
hypersurface of the hypersphere, draw the straight line NP in 4-D
space and continue it until it hits a unique point P' of your E3. It is

important here to realize that it is possible for a line in 4-D space to meet a 3-D space in just one point.

Let's get back to something a little less mind-boggling: A. Square. It was clear to him that his space was spherical. After all, Livingchip had traveled "around space," had he not? Surprisingly, or perhaps not so surprisingly, A. Square's theory was universally rejected. "Space cannot be curved, Professor Square," his boss told him with a trace of the guillotine in his voice; "space by its very nature is flat. God would not create an imperfect universe." Square replied, "But can't you see? Our space is curved in the direction of the third dimension, how else could Livingchip have journeyed around the universe without ever curving left or right?" His boss snapped, "Doctor Square, the third dimension is not *real*. It is only a metaphor for that which is miraculous and intrinsically inexplicable. And as for Livingchip's trick . . . our Father Twistor is working on that small anomaly."

Father Twistor was head preacher at the Church of the Third Dimension. He had founded the Church during the turbulent times following A. Square's escape from the guillotine. Confused and frightened by this incursion of "extraordinary reality" (to use Father Twistor's phrase) upon their lives, the Flatlanders had cast about for a leader to make their changed world intelligible to them, and Father Twistor gave them what they wanted. A. Square could easily have seized power, but his trips into the third dimension and his months in jail had soured him on the polygonal race. He was content to lead a relatively isolated life at Flatland U. It was not hard to see him, but few took the effort. He was, after all, something of an eccentric.

Father Twistor was a good and ingenious mathematician, but he had a fundamental disbelief in the third dimension. He was masterful at finding 2-D explanations for 3-D phenomena, while paying lip service to the third dimension. He used the words "three-dimensional" and "miraculous" interchangeably, and was not above passing off cheap magic tricks as "three-dimensional phenomena." The Church of the Third Dimension was a great success because it made something comfortably "miraculous" out of events that had initially been uncomfortably real.

Soon after A. Square's conversation with his boss, Father Twistor came to see him. "Well, Professor," Twistor began heartily, "up to your old tricks? What's this I hear about a spherical space? Leave the third dimension to the theologians! If there's any third dimension in the *real* world, it's time; there's no third space dimension to bend things in!"

"All right Twistor," Square answered, "you must have some

miraculous 2-D explanation up your sleeve. Let's hear it."

"There's nothing to it," Twistor answered expansively. "Livingchip *grew* as he moved away from Flatsburg. He grew so fast that he reached Infinity after a year. There's only one Infinity, so he was able to come back from it in any direction he liked. He happened to pick it so he came back from the west even though he left toward the east." Twistor beamed soothingly at A. Square's furious countenance.

"That's ridiculous!" A. Square cried.

"Not ridiculous, dear Square, miraculous," Twistor responded.

Let's take a look at Father Twistor's idea. A. Square is thinking in terms of the real sphere and Father Twistor is thinking in terms of the Flat Sphere. Since the two spaces are isomorphic, I can tell you right now that no one is going to win the argument. And this is going to be our point: curvature of space in a higher dimension can be explained away if you assume that objects stretch and shrink in the right way as they are moved around your idealized flat space. Figure 64 illustrates Livingchip's journey the way Father Twistor saw it.

Notice that if you take a spherical Flatland and put a plane touching it at Flatsburg, then the stereographic-projection image of a square moving around the sphere as illustrated in Figure 57 looks just the way Father Twistor says. When the square contains the point from which the projection lines are drawn, then its image is infinite. This makes sense, for if the residents of Flatsburg had looked through a powerful telescope at the time when Livingchip was at the point on Flatland's spherical space diametrically opposite from Flatsburg—if they had looked in any direction at *all*—then they would have seen Livingchip. Now if we were to see a part of some person dimly in the background, no matter which way we pointed our telescopes, then we would conclude that this "person" was infinitely big. This would happen, for instance, if an astronaut was floating at the point in space most distant from us, assuming that our 3-D space is spherical. An odd feature of this astronaut's appearance, which is apparent from Figure 64, is that he would be "inside out"; that is, instead of his skin forming a surface on the inside of which were his innards and on the outside of which was us, his skin would be a surface on the *outside* of which was his innards and on the *inside* of which was us. Would he notice anything strange? No! He would feel perfectly normal. Only *we* would look infinitely large and inside out to *him*. This weird behavior of the astronaut's body at "infinity" is what is called a "coordinate singularity of space" as opposed to an "essential singularity of space." That is, it is strange behavior of space which is only apparent, and which can be eliminated by looking at things in a different way.

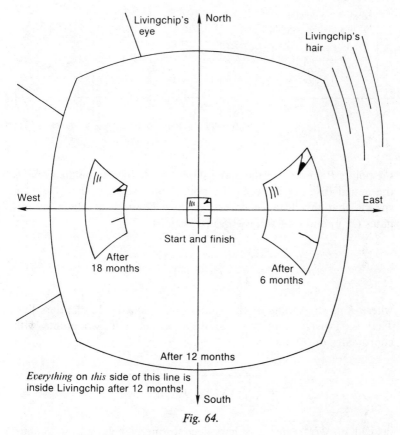

Fig. 64.

Getting back to the Square vs. Twistor debate, it turns out that Twistor had actually worked out a formula for how fast Livingchip had to grow. The idea was that the space of Flatland was an infinite Cartesian plane—every point has an (x, y) coordinate, and we take the point $(0, 0)$ to be Flatsburg—but that the change in distance ds between the two points with coordinates (x, y) and $(x + dx, y + dy)$ was not going to simply be the square root of $dx^2 + dy^2$, as it would be if everything was normal. The idea is that, given the two points (x, y) and $(x + dx, y + dy)$ in their space, it is not absolutely necessary that the Flatlanders assume that the "distance," or amount of space, between these two points is automatically going to be the square root of $dx^2 + dy^2$, as the Pythagorean Theorem would suggest. Assuming the Pythagorean theorem is, after all, equivalent to assuming Euclid's Fifth Postulate, as was pointed out in Chapter 2. Perhaps the plane on which the Flatlanders live was selectively stretched and shrunk *after* each point had been assigned its Cartesian coordinates.

It is possible to calculate, by assuming that the distance between

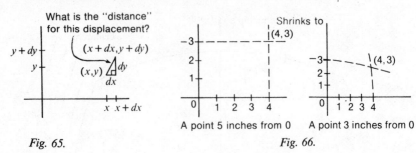

Fig. 65.   Fig. 66.

the points P' and Q' in the plane should be defined to be the same as
the actual distance between the points' preimages P and Q on the
sphere, that the distance ds between the points with plane coordinates (x, y) and $(x + dx, y + dy)$ is given by

$$ds = \frac{1}{1 + \frac{1}{4K^2}(x^2 + y^2)} \sqrt{dx^2 + dy^2} \, ,$$

where K is the radius of the sphere on which the Flatlanders live.
That is, Twistor said, the real distance between two points with
coordinates (x, y) and $(x + dx, y + dy)$ is

$$\frac{1}{1 + \frac{1}{4K^2}(x^2 + y^2)} \sqrt{dx^2 + dy^2}$$

instead of $\sqrt{dx^2 + dy^2}$, as had been formerly believed. A. Square
would view K as being the radius of the universe, but Twistor,
believing in flat space, would think of K as rather some sort of
universal constant with no necessary physical correlate. We can view
a plane as a sphere with infinite radius and observe that if K is
infinite, Twistor's formula for distance reduces to the ordinary distance formula.

We should pause here and mention what is meant exactly by dx
and dy. The terms dx and dy are understood to be *infinitesimals*,
non-zero quantities that are smaller in absolute value than any real
number. Of what use is a formula for the infinitesimal distance
between two infinitesimally close points, you may ask. The idea is
that we have in the calculus a tool for adding together infinitely
many infinitesimals to get ordinary real numbers. This process is
called integration. The distance between two points P' and Q' along
a given line m is defined to be the infinite sum of all the infinitesimal
distance elements along the line m between P' and Q' (usually

written $\int_P^Q ds$), where it is normally assumed that you know which line *m* you have mind. Thus the distance from Flatsburg (0, 0) to the eastward edge of the universe (∞, 0) is

$$\int_{(0,\, 0)}^{(\infty,\, 0)} ds = \int_{x=0}^{x=\infty} \frac{1}{1 + \dfrac{1}{4K^2}\, x^2}\, dx$$

$$= 2K \arc \tan \frac{x}{2K}\bigg]_0^\infty = \pi K,$$

where we have used the fact that our formula for *ds* gives the distance between (*x*, 0) and (*x* + *dx*, 0) to be $\dfrac{1}{1 + \left(\dfrac{x}{2K}\right)^2}\, dx$. Notice here that the value πK is exactly the distance from a point on a sphere of radius *K* to a point on the opposite side of the sphere.

What is the relationship between Twistor's claim that the Flatlanders grow larger as they move away from the origin and the formula for *ds* which says that the greater $x^2 + y^2$ is, the smaller will be the distance change *ds* associated with a given coordinate change *dx*, *dy*? The two ideas are essentially equivalent. Say that you have an *x*-axis on which you have labeled certain points 0, 1, 2 and so on. Now say that you have a rod at rest on the *x*-axis with its left end at the point 0 and its right end at the point 1. Now say that you slide the rod to the right and find that it comes to rest with its left end at the point 2 and its right end at the point 4. There are two possible conclusions you can draw: (a) As you moved the rod to the right it expanded from one unit in length to two units in length, or (b) the actual distance between the points 2 and 4 is the same as the actual distance between the points 0 and 1 (Figure 67).

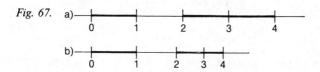

Fig. 67.

Conclusion (a) seems natural if you believe that we have a fixed underlying Euclidean space with nice Cartesian coordinates for each point. When apparently non-Euclidean phenomena arise in your world, you explain that matter is subject to strange contractions and expansions due to its position in space and insist that your underlying space is none the less Euclidean.

Conclusion (b) seems correct if you feel that a rigid body such as a ruler should not be viewed as stretching or shrinking according to where it is located in space. The feeling here is that if you pick up a yardstick and carry it over to the other side of the galaxy, it is still going to be one yard long. When apparently non-Euclidean phenomena arise in your world, you explain that your space is in fact *not* Euclidean, which is why any attempt to lay out Cartesian coordinates satisfying $ds^2 = dx^2 + dy^2$ ends in failure.

The modern tendency is to take conclusion (b), arguing that since the absolute space of conclusion (a) is unobservable it has no real existence. The modern approach is to lay down coordinate lines in any natural fashion and only then to bring in distance by actually measuring at each point (x, y) the distance associated with a given coordinate displacement. It turns out that if we assume that space is locally flat (this means that in any small enough region, space appears Euclidean) then there will be three functions $g_{11}(x, y)$, $g_{12}(x, y)$ and $g_{22}(x, y)$ of position such that

$$ds^2 = g_{11}(x, y)\, dx^2 + 2g_{12}(x, y)\, dx\, dy + g_{22}(x, y)\, dy^2.$$

Often the three g-functions are combined into one function

$$G(x, y) = \begin{bmatrix} g_{11}(x, y) & g_{12}(x, y) \\ g_{12}(x, y) & g_{22}(x, y) \end{bmatrix}$$

whose value at each position is a matrix, or tensor. In the 3-D case we have a similar function $G(x, y, z)$ equal to the symmetrical matrix

$$G(x, y, z) = \begin{bmatrix} g_{11}(x, y, z) & g_{12}(x, y, z) & g_{13}(x, y, z) \\ g_{12}(x, y, z) & g_{22}(x, y, z) & g_{23}(x, y, z) \\ g_{13}(x, y, z) & g_{23}(x, y, z) & g_{33}(x, y, z) \end{bmatrix},$$

where

$$ds^2 = g_{11}(x, y, z)\, dx^2 + 2g_{12}(x, y, z)\, dx\, dy + 2g_{13}(x, y, z)\, dx\, dy \\ + g_{22}(x, y, z)\, dy^2 + 2g_{23}(x, y, z)\, dy\, dz \\ + g_{33}(x, y, z)\, dz^2.$$

The *G*-function is called the *metric tensor*. It turns out that if you are given an arbitrary coordinatization of space and the metric tensor at each coordinate point, then you know all there is to know about the structure of space. If you had laid out your coordinates differently, you would have obtained a different metric tensor, but the new metric tensor would be related to the old one in a certain natural way.

As we saw before, if you take the regular Cartesian coordinates on the plane then

$$G(x,y) = \begin{bmatrix} \left(\dfrac{1}{1 + \dfrac{x^2 + y^2}{4K^2}} \right)^2 & 0 \\ 0 & \left(\dfrac{1}{1 + \dfrac{x^2 + y^2}{4K^2}} \right)^2 \end{bmatrix}$$

leads to spherical space; that is, it makes the plane look exactly like the Flat Sphere.

Let us clarify this. Why, if $G(x,y)$ is as we just specified, should our shortest paths look like the circles in Figure 43?

Or, to put it back in an *ad polygonem* form, how could Father Twistor argue that his supposed expansion of someone moving away from Flatsburg would cause the shortest path between P and Q to be the curved line rather than the straight line (Figure 68)? The answer is simple. Since a ruler will get longer as we move it away from 0, if we lay the ruler down repeatedly along the curved path PQ, we will have to lay it down less often then we would if we laid it down repeatedly along the straight path PQ. A string stretched from P to Q would actually go along the curved path. Notice that the triangle OPQ is a right triangle with an angle sum of 270°.

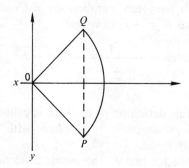

Fig. 68.

Now we can begin to see how to imagine our 3-D space to be spherical. Granted that we can't readily imagine curvature in the direction of the fourth dimension, we *can* imagine stretching rulers. So the idea is to start with a mental image of nice ordinary 3-D Euclidean space. Now assume that as any ruler or other object moves away from the origin it expands. The expansion is determined by the formula

$$ds = \frac{1}{1 + \dfrac{x^2 + y^2 + z^2}{4K^2}} \sqrt{dx^2 + dy^2 + dz^2} \ ,$$

where K is again the desired radius of the hypersphere whose hypersurface we are to occupy. What this formula says is that the distance change produced by a coordinate change (dx, dy, dz) is

$$1/1 + \frac{x^2 + y^2 + z^2}{4K^2}$$

as big at the point (x, y, z) as it would have been at the origin. Put differently, it requires

$$1 + \frac{x^2 + y^2 + z^2}{4K^2}$$

times as big a coordinate change at the point (x, y, z) to produce the same distance change as such a coordinate change would have produced at the origin. Again, this means that if we move an object of a certain fixed size from the origin to the point (x, y, z), then the size of the object relative to our coordinate system must increase. In particular, the size of the object must grow rapidly enough for it to be able to reach and pass through infinity after a finite amount of time.

An interesting situation arises if we take the radius of our universe to be an *imaginary* number, say i. Consider the 2-D case first. If $K = i$, then $K^2 = -1$, so we get

$$ds = \frac{1}{1 - \dfrac{x^2 + y^2}{4}} \sqrt{dx^2 + dy^2}$$

The plane with this definition of distance is called the Flat Pseudosphere, where a *pseudosphere* is a sphere with imaginary radius, whatever that may mean.

As in the case for real K, the metric distortion at the origin is nonexistent, that is, at $(0, 0)$, $ds = dx^2 + dy^2$. What happens here as we move away from the origin? As we get closer to the circle around the origin of radius 2, $x^2 + y^2$ gets closer to 4, and the denominator of our expression for ds approaches zero. This means that as you get close to the indicated circle, ds gets to be very large for even small (dx, dy). This can be envisioned by imagining a yardstick that shrinks as it approaches the circle of radius 2, shrinks so rapidly that the distance between the points 0 and 2 in Figure 69 is infinite!

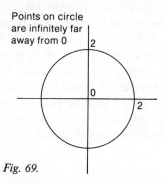

Points on circle
are infinitely far
away from 0

Fig. 69.

Things get even weirder *outside* the circle. Here all the ds are *negative*. The distance between any two points outside the circle is a negative number. Moreover, if P is a point outside the circle, then the distance from P to infinity is some finite negative number, while the distance from P to any point on the circumference of the circle is—infinity.

Before we start trying to figure out what the geodesics look like in this situation, let me tell you that there is no curved surface (in 3-D Euclidean space) that corresponds to the Flat Pseudosphere in the same way that the sphere corresponds to the Flat Sphere. (This was proved by David Hilbert in 1901.) The notion, due to B. Riemann, of representing a surface by the Euclidean plane with a tensor-valued function determining the form of ds is essentially richer than the notion of representing a surface by a deformed plane in 3-D space. There is no real Pseudosphere in our 3-D space, but we are able to represent it analytically by the above formula for ds.

Let us restrict our attention to the region of the plane *inside* the circle around the origin of radius 2. With the pseudospherical metric this region appears to be infinite, although it is finite under the regular Euclidean metric. This is as opposed to the spherical metric mentioned before, which makes the entire plane appear to be a finite region, although the plane is infinite under the regular Euclidean metric.

Just as we represented spherical 3-D space by thinking of Euclidean 3-D space in which our rulers grow as they move away from the origin, we can represent a pseudospherical 3-D space by thinking of the inside of a sphere in regular Euclidean space in which rulers *shrink* as they move away from the center and toward the surface of the sphere. If the rulers shrink fast enough, the distance from the center of the sphere to its surface will appear infinite. Thus you could have a universe which went on infinitely in every direction as far as you could tell, but which was really just the 3-D space inside a tennis ball. It's just a matter of having everything shrink as it moves away from the center of the ball. It's like the old paradox that you can never leave the room you're in because you have to go half the distance, then half the remaining distance, then half the remaining distance, ad infinitum (Figure 70). But if every time you went one of those halves you shrank by a factor of 1/2, then each one of those infinitely many steps would be the same distance, say three feet, for you. And you really *couldn't* get out of the room.

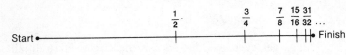

Fig. 70.

The pseudospherical space is negatively curved, as opposed to the spherical space, which was positively curved. Going back to the plane version, what are the geodesics like? Let's just think about the part of the plane inside the circle of radius 2 with the distance

$$ds = \frac{1}{1 - \dfrac{x^2 + y^2}{4}} \sqrt{dx^2 + dy^2}$$

as before. It turns out that a geodesic is an arc of a circle that cuts the circle of radius 2 at right angles.

Figure 71 (due to Poincaré) is one of the best models of hyperbolic geometry that we have. Notice, for instance, that it is possible to have many lines passing through a given point P which do not intersect a given line m.

Earlier we discussed the way in which a Flatlander on a sphere would return to his starting point if he traveled long enough in any direction. Let us now consider a different sort of Flatland universe in which he would return ... but as his mirror image.

The surface depicted in Figure 72 is called a Möbius strip. You can easily make one by taking a strip of paper and joining it so the

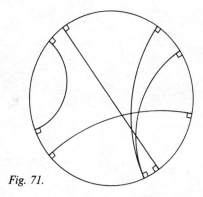

Fig. 71.

two ends have opposite orientation, in the sense indicated in Figure 73.

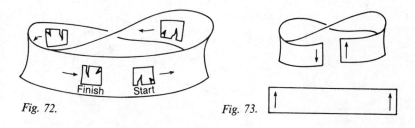

Fig. 72.   Fig. 73.

Notice that when A. Square slides around the Möbius strip he is indeed turned into his mirror image. One might be tempted to say that when he gets back he is "on the other side" of the sheet of space. But if his space is really 2-D, there is no such thing as being on one side or the other. A way of visualizing this in terms of a paper Möbius strip would be to imagine that A. Square is drawn with ink that soaks through the paper.

It would, of course, be unnatural for Flatland to be the surface of a Möbius strip. The strip has edges, and no one's space should have edges. We could make the edges unreachable by postulating a field that caused any Flatlander to shrink and shrink when he moved toward an edge. But there is a better way.

The surface shown in Figure 74 is a Klein bottle.* It is constructed, in theory, by taking a cylinder and joining the two ends so that they have opposite orientation (Figure 75).

One thing wrong with our attempts to construct a Klein bottle in

---

*The Klein bottle pictures are from D. Hilbert and S. Cohn-Vossen, *Geometry and the Imagination* (Chelsea Publishing Co., N.Y., 1952), p. 308.

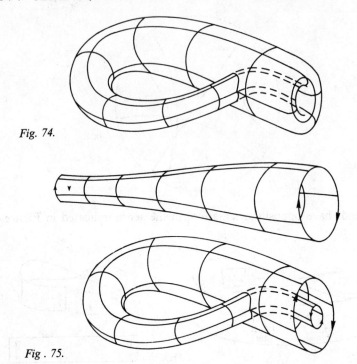

Fig. 74.

Fig . 75.

3-D space is that it is necessary to have this surface intersect itself. One must pretend that an object moving in the surface is free to move through the "wall" where the bottle penetrates itself.

In 4-D space it would be possible to make a perfect Klein bottle. To see this, imagine the Flatlanders trying to make a Möbius strip. They would have to proceed somewhat as in Figure 76 in order to join two ends of a strip with opposite orientation. Of course, when there are three dimensions to work in, you take the left end of the strip out of the plane of the right end, turn it over and come back into the plane to join up.

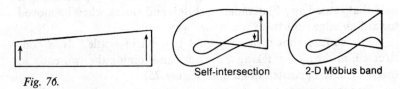

Self-intersection   2-D Möbius band

Fig. 76.

In the same way, in 4-D space, we could move the left end out of the space of the right end, "turn it over" and then come back into the space of the right end to join up. You can try to visualize the smooth

Klein bottle in 4-D space by thinking of pulling at the loop of the 3-D Klein bottle until the region of self-intersection has been eliminated.

Note that if A. Square lived on a Klein bottle, then traveling in certain directions would bring him back to his starting point without mirror reversal, but traveling in certain other directions *would* bring him back as his mirror image. And this is how it would be if our space was curved into the hypersurface of a Klein hyperbottle in 5-D space.

PROBLEMS ON CHAPTER 3

(1) Assume that galaxies are uniformly distributed in space. If our space were Euclidean, then for any r, the number of galaxies within a distance of r of our galaxy would be r^3 times some fixed constant. Would this be the case if our space were the hypersurface of a hypersphere?

(2) Suppose that our space is hyperspherical, and that a fleet of space ships flies directly away from Earth in many different directions. Where will these ships first meet again?

(3) Imagine a spherical mirror (like a Christmas-tree ornament) and imagine that the entire universe outside the mirror is reflected inside it. As a person moves away from the mirror toward infinity, his image moves toward the center of the mirror, shrinking all the while. Would you notice if you actually lived in the mirror world?

(4) It is sometimes said that an object shrinks when it enters a strong gravitational field (e.g., near the surface of the sun). Can you think of a way of expressing this fact which allows you to say that objects do not actually shrink or expand as they move about?

(5) It has been suggested that mirror-reversed matter would be what is known as antimatter. It is easy to notice antimatter since it combines with normal matter to explode. Now recall that going around the Klein bottle in one direction would cause Flatland matter to turn into its mirror image, but going around it in another direction would not. If we were to observe that antimatter fell on the Earth from only certain directions in space, what might we hypothesize?

(6) Imagine Flatland to be a marble surface which has a heat source at point 0. Say that the Flatlanders and their measuring instruments all undergo expansion when they are heated. Will A. Square view the space of Flatland as being positively, or negatively, curved in the vicinity of the point 0?

4

TIME AS A HIGHER DIMENSION

As I write this there is a fly zooming around my desk. It's almost winter and he has come in to get warm and eat garbage. When he is in motion (and now that I am writing about him he is putting on quite a show) I do not actually see a moving black object. I see, rather, a sort of trail in space (Figure 77).

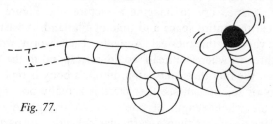

Fig. 77.

Pause here and wave your right hand in a complicated 3-D pattern. Look at the trails. In what sense do they exist? What would it be like if your hand was at each of its positions at once? What if you move your hand from your nose to your ear and then back to your nose—why doesn't the old hand at the nose get in the way of the new hand at the nose?

The viewpoint we wish to develop in this chapter is that all 3-D objects are actually trails in 4-D space-time. "Space" is a fairly arbitrary 3-D cross section of space-time which we imagine to be moving forward in the direction of the remaining dimension, "time."

Is, then, time the fourth dimension? Not necessarily, You could still have four dimensions—say three to live in and one to curve space in the direction of—and then throw in time as the fifth dimension. It is possible and useful to view time as a higher dimension, but the reader should not jump to the conclusion that whenever we talk about a higher dimension we are referring to time; many of the ideas about the fourth dimension that we have outlined are no longer valid if you insist that the fourth dimension is simply time. Some things that are possible in pure four-dimensional space are not possible in four-dimensional space-time.

To get a good mental image of space-time, let us return to Flatland. Suppose that A. Square is sitting alone in a field. At noon he sees his father, A. Triangle, approaching from the west. A. Triangle reaches A. Square's side at 12:05, talks to him briefly, and then slides back to where he came from. Now, if we think of time as being a direction perpendicular to space, then we can represent the Flatlanders' time as a direction perpendicular to the plane of Flatland. Assuming that "later in time" and "higher in the third dimension" are the same thing, we can represent a motionless Flatlander by a vertical worm or trail and a moving Flatlander by a curving worm or trail, as we have done in Figure 78.

We can think of these 3-D space-time worms as existing timelessly. We can use them to produce an animated Flatland by taking a 2-D plane, moving it upward (forward in time) and watching the motions of the figures formed by the intersections of the worms with the moving plane. Try to imagine a picture like Figure 78 which encompassed the entire space and time of Flatland. A vast tangle of worms of varying thicknesses! Actually, each worm would be a tangle of threads, where a thread would correspond to the trail of an atom. Given the fact that every atom in one's body is replaced every seven years or so, we can see that there is actually no single thread that goes the whole length of one's life. A living individual is a persistent *pattern* rather than a particular collection of particles.

It is an interesting mental exercise to try to see our world in terms of space-time. Walking through a crowd of people, for instance, one can try to see the people as trails in space-time rather than as spatial objects moving forward in space-time. Under this view what our world really consists of is "worms" in 4-D space-time. The universe at any instant is a particular 3-D cross section of this 4-D structure.

A question that arises if we attempt to accept the view that our universe is a static space-time configuration is, "Why can't we *see* the past and the future if they really exist? What causes us to perceive oursel moving forward in time?" In other words, if we take the two worms in Figure 78 and let them exist statically, this does not seem to provide for A. Square's feeling of moving forward in time. One might suggest that we take the static space-time worms and

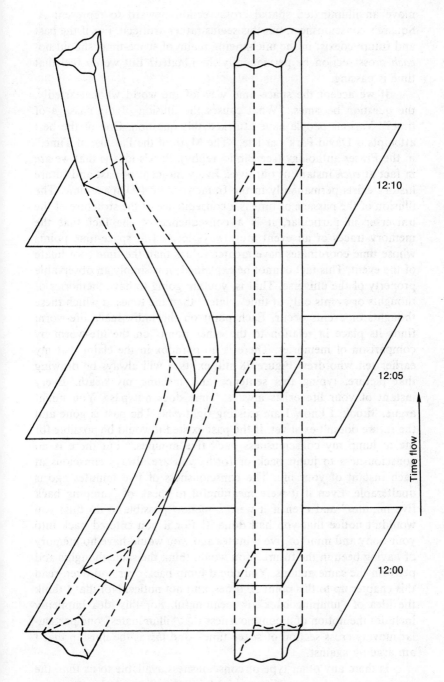

Fig. 78.

move an illuminated spatial cross section upward to represent A. Square's consciousness, but this seems rather artificial. For if the past and future coexist in the unchanging realm of space-time, should not *each* cross section be permanently illuminated? But we *do* feel that time is passing.

If we accept the space-time view of the world wholeheartedly, the question becomes, "What causes the illusion of the passage of time?" Various people have attacked this question. One of the best attempts is David Park's article, "The Myth of the Passage of Time," in the Fraser anthology (see Bibliography). Park's idea is that we are in fact at *each* instant of our lives. Every moment of past and future history exists permanently in the framework of 4-D space-time. The illusion of the passage of time is a consequence of the structure of the universe; in particular, it is a consequence of the fact that the memory traces of an event are always located at space-time points whose time coordinates have greater values than the time coordinate of the event. This fact cannot be explained; it is simply an observable property of the universe. That is, you are going to have memories of thoughts or events only at times "later" than the times at which these thoughts or events occur. Each point on the individual's life-worm finds its place in relation to the other points on the life-worm by comparison of memories. There is no paradox in the claim that my earlier self who drew Figure 78 still exists. I will always be drawing that picture, typing this sentence and meeting my death. Every instant of your life exists always. Time does not pass. You might argue, "Look, I know I am existing right *now*. The past is gone and the future doesn't exist yet. If the past existed it would be possible for me to jump my consciousness back five minutes." But there is no consciousness to jump back or forth; you are always conscious at each instant of your life. The consciousness of five minutes ago is unalterable. Even if it were meaningful to speak of "jumping back five minutes" and even if it were somehow possible to do this; you wouldn't notice that you had done it! For if you entered back into your body and mind of five minutes ago, you would have no memory of having been in the future. You would think the same thoughts and perform the same actions. You could jump back over and over, read this chapter up to this point 50 times, and not notice. Not that I think the idea of "jumping back" *is* meaningful. For this idea implicitly includes the notion of a consciousness that "illuminates" one particular moving cross section of space-time—and *this* is the illusion that I am arguing against.

Is there any other type of consciousness available to us than the various points along our life-worm? Is there any way to be conscious in static space-time instead of in moving space? Such consciousness is the goal of the mystic's quest. Practitioners of Yoga speak of immortality and freedom, consciousness of the Eternal Now, and the transcendence of time. Are they talking about the direct perception of the unchanging world of space-time? Reports indicate that a period of time spent in deep meditation is recalled as an essentially timeless period. It seems to be in fact possible to escape the feeling of time passing. The actual experience of such "illumination" cannot ordinarily be fully recalled in ordinary states of consciousness. Does this mean that such states of consciousness supply a direct window into the world of space-time? Perhaps, but perhaps not. It can be argued that the production of a "timeless" feeling is simply a trick that works as follows. The way in which we notice that time is passing is that each instant of consciousness is different from the ones just before and just after it. This is because we are always thinking new thoughts, noticing new things. Now, the technique by which one enters a yogic trance is to *stop* thinking new thoughts. This is done either by thinking nothing at all (this is not easy!) or by concentrating one's attention on a repetitive thought loop (e.g., a mantra such as the currently popular "Nam myoho renge kyo" or the "Om mane padme hum" mantra of Baba Ram Dass, author of the excellent introduction to Yoga, *Be Here Now*). Now if you are thinking nothing, there is no way to differentiate one instant from the last instant or the next instant. If you are doing mantra there is no way to differentiate one repetition from the last repetition or the next repetition. Hence either of these mental exercises leads to a sensation of timelessness (Figure 79). Let me interject here that there is nothing peculiarily Eastern about the use of mantra; the "Hail Mary" is

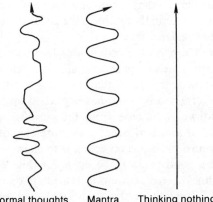

Fig. 79.   Normal thoughts   Mantra   Thinking nothing

perhaps the most widely used Western mantra. Now, these sensations of timelessness are pleasant and valuable, but are they really 4-D consciousness? Perhaps not, but they are a good first place to start one's efforts to develop a 4-D consciousness.

A different trick for developing a space-time consciousness is described in Carlos Castaneda's book, *A Separate Reality*, an account of a Mexican Indian named Don Juan and his attempts to teach or show Castaneda a new way of interpreting reality. Certain sequences in the book give one the impression that Don Juan was actually trying to teach Castaneda to see in space-time. One of the exercises which Don Juan assigned was that Castaneda should start paying attention to sounds instead of to sights. This may sound unimportant, but civilized man is in fact highly visually oriented. Most of our information (e.g., the printed word) comes to us through our eyes, as opposed, say, to a primitive hunter who depends to a much larger degree on his ears (e.g., tribal chants and sounds of animals). The interesting thing about our ears is that they perceive *time* structure instead of *space* structure. In other words, you can't hear what's going on in a room with a "glance" of your ears. It takes time to hear what's going on. Notice, for instance the way in which you hear a song on the radio. You do not hear it a note at a time. You hear chords, progressions, crescendos and so on. You perceive time-structure.

Viewing events in a historical perspective is another way to get closer to a space-time world view. That is, you can become more aware of yourself as a space-time structure if you keep in mind the way you were five minutes, five hours, five years ago. There are even moments of intense recollection when we actually seem to go back to the scene of a past event. The Argentinian writer Jorge Luis Borges goes so far as to argue, in his paradoxically entitled essay, "A New Refutation of Time," that when you recreate a particular state of consciousness you actually return to the time when that original state of consciousness existed in you.

Let us now discuss the problem of free will. A common objection to the view that all space and all time can be rolled into one static space-time structure is that the future does *not* seem to be completely determined by what has happened up to this instant. The feeling is that we *do* choose from the various possible courses of action open to us and that hence the future cannot already exist.

The easy answer to this objection is to claim that we do *not* have free will, and a good case can be made for this. Whenever someone performs an unexpected action our immediate question is, "Why did you do *that?*" Implicit in this question is our belief that there *is* always a reason for a person's action, that, in fact, he does *not* have free will but responds only to the forces of internal and external pressures.

This answer is not entirely satisfactory since there do seem to be choices that are completely nonpredictable. Little choices, such as which shoe to put on first, seem to be made in a random and nonpredictable fashion. In physics we have events that seem to be *fundamentally* random. If you have, for instance, an atom of uranium there is *even in principle* no way to predict if it will decay and emit an alpha particle within the next ten seconds or not. How can this already be decided in space-time if we can't predict it?

Well, why not? After all, predetermination does not imply predictability. All of the future could already exist, including the unpredictable little zigs and zags that occur. Still, there is something a little unsatisfying about this state of affairs. The feeling is that if there is nothing forcing the atom of uranium to decay or not to decay in the next ten seconds, then it should be possible for it to do either one. But if the future already exists, then it isn't *really* possible for it to do either one. Either it's going to decay, in which case it wasn't actually possible for it not to decay, or it isn't going to decay, in which case it wasn't actually possible for it to decay, although we didn't know this. Is there some way to set up the universe so that the different possible futures are *real* possibilities instead of *theoretical* possibilities?

Yes, there is. The idea is that we can work with a *branching* universe. This idea has been seriously proposed by several physicists (see DeWitt, ed., *The Many-Worlds Interpretation of Quantum Mechanics*). To get a picture of it let's work with a zero-dimensional space: Pointland, a space consisting of one point. Now say that this point can decide at the end of each second whether or not to glow during the following second. Now, if we draw the life-worm of this point we get a line going upward (forward in time) that will be lit up during some one-second intervals and dark during some one-second intervals (Figure 80). Since this whole line exists in space-time, we might conclude that the point's feeling that it was deciding at the end of each second whether or not to glow for the next second was illusory.

In order for the point's choices to be *real*, it is necessary that its world line split in two each time it makes his glow/no-glow decision (Figure 81). That is, all of its possible futures really exist. It will have the illusion that it only experiences one of them, but in fact there are many of it, experiencing every possible life. Each one of these "selves" will have the illusion that it is unique, will have the illusion that it has judiciously selected a particular sequence of glow/no-

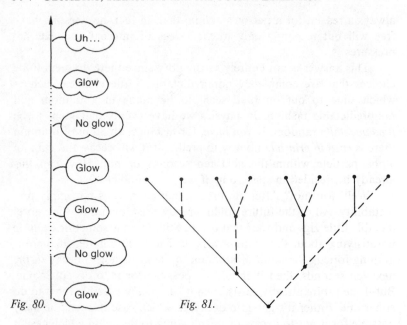

Fig. 80.   *Fig. 81.*

glows, will feel that its free will has realized only one of the many possible universes. In fact, all the possible universes will exist.

If we accept this "branching-universe model" for our own universe we can see that there will be a staggering number of branches to our universe. For every time some indeterminate quantum event does or does not take place in an atom, the universe splits into two branches. That's a lot of new branches per second! Would every possible universe exist, then? Would there be, say, a universe in which you were Superman? Sure, in order for you to fly it would just be necessary that all the atoms in your body be coincidentally moving upwards at the same time in the course of their random fluctuations. Unlikely, but not impossible! To take a more realistic example, consider the paradox called "Schrödinger's Cat," described by Erwin Schrödinger, one of the founders of quantum mechanics.

A cat is left in a room with a closed glass bottle of cyanide gas. Next to the bottle is a hammer connected to a Geiger counter, which is next to a small amount of uranium. The hammer is coupled to the Geiger counter in such a way that if an atom of the uranium decays between 12 noon and 12:01 P.M., the Geiger counter will sense this and cause the hammer to smash the bottle, thus killing the cat. The paradox is that until we return to the room, say around 6 in the evening, and observe whether or not the cat is alive, it is not physically meaningful (according to quantum mechanics) to say that the cat is definitely dead or definitely alive. There is a certain probability that an atom decayed during the crucial minute after noon, and until we make an observation as to what actually happened, both possible worlds have a certain theoretical existence. The uncertainty arises because the laws of quantum mechanics merely describe the evolution of certain probabilities with the passage of time; the precipitation of a particular observation out of the probability space of quantum mechanics is not a phenomenon that can be accounted for in any deterministic way.

Hugh Everett's solution to this situation (his paper appears with commentaries in the DeWitt book) is to maintain that every state in the probability space of quantum mechanics really exists: there is a universe in which the cat lives and a universe in which the cat dies, and we split and enter both universes. How many dimensions would we need for such a branching universe?

In one sense it seems that we'd only need five: three for space, one for time and one in the direction of which the universes could do their branching. On the other hand, if we think of the branching induced by any one particle as being independent of the branching induced by any other particle, we'd want a dimension for each particle in the universe—which is a lot of dimensions.

We discussed the idea of developing a space-time consciousness when we started viewing time as a static dimension. Is there any chance of being able to somehow sense all the different possible universes, assuming with Everett that they "really" exist? Maybe we are, in some way, aware of many possible worlds, and we shift our attention back and forth from one to the other. One day everyone loves you, the next everyone hates you; one minute everything is Love, the next it's curved space-time; you see the blue sky shining through the trees, blink and you see the green leaves in front of the sky. No less a man than Ludwig Wittgenstein has said, "The pessimist and the optimist live in different worlds"; why not take this literally? Assuming that we have a sort of access to many possible universes, what should we do to know them all? That is, assuming that the *true* reality is composed of the many possible *individual* realities, what can we do to tune in on the whole big thing instead of the particular channels? It would be a matter of stopping the internal process of naming, evaluating, judging, discriminating and so on that is involved in the forming of world views. The only way not to be tied to a particular system of interpretation is to have none. In the words of Don Juan (from page 264 of Castaneda's *A Separate Reality*), "The world is such-and-such or so-and-so only because we tell ourselves that that is the way it is. If we stop telling ourselves that the world is so-and-so, the world will stop being so-and-so. At this moment I don't think you're ready for such a momentous blow, therefore you must start slowly to undo the world."

What is reality? NO MIND! None the less, getting there is half the fun.

PROBLEMS ON CHAPTER 4

(1) If you say that the fourth dimension is time, then it is possible for you to construct a hypersphere in space and time. How?

(2) Space-time is not really the same as 4-D space for a number of reasons. For instance, the ability to move backward and forward in time *would* enable you to get into a sealed room (how?), but it would *not* enable you to remove someone's supper from his stomach without disturbing him (why not?).

(3) Kurt Vonnegut's novel *Slaughterhouse Five* is about a guy who lives his life in a jumbled order; for instance, first he experiences 1950, then 1946, then 1956, then 1943, etc. Would you *necessarily* notice if you lived your life in this manner? Is it meaningful to claim that you have done so? In the Vonnegut book the character *does* notice his jumping about because his memory is continuous. That is, in 1946 he remembers 1950, etc. In what way is this state of affairs incompatible with the space-time view I argued for in this chapter?

(4) If the time of our universe really branches, is there any way in which you can influence which branch you go into? Is this a meaningful question? People sometimes throw coins to get a hexagram which they look up in the *I Ching* to find out which branch of the universe they are entering. Could one improve his world by getting good at throwing the *I Ching*?

(5) In quantum mechanics a system (e.g., a person) is represented by a "state vector" in Hilbert space that codes up the extent to which it or he is in each of the many possible universes. A system's state vector looks something like this: $\langle \frac{1}{10}, \frac{1}{2}, \frac{1}{4}, \frac{1}{20}, \ldots \rangle$, where the sum of the entries is 1 and each entry indicates the probability that a measurement (e.g., of position) will find the system in the state corresponding to that slot. What would a system's state vector look

like right after a measurement is performed, and you have forced the system to be in only one universe?

(6) It can be maintained that we are justified in saying that event A happens before event B only if at event B we have some evidence (e.g., memory) that event A has taken place. Would, then, your thoughts necessarily be linearly ordered in time?

5

SPECIAL RELATIVITY

In the first part of the last chapter I discussed the idea that the 3-D world we live in at any moment is but a cross section of 4-D space-time. Given the way that things are in our world, what can we infer about the structure of space-time? What is the geometry of space-time? What type of metric does it have?

In 1905 Albert Einstein first gave serious consideration to these questions in his paper, "On the Electrodynamics of Moving Bodies." This is the paper in which his celebrated Special Theory of Relativity was first presented. The paper is rather analytical and contains no pictures. In 1908, H. Minkowski, a young Russian mathematician, presented a paper in which he interpreted Special Relativity as a theory about the geometry of space-time. The paper, called "Space and Time," introduces a type of picture called a Minkowski diagram. Let me quote the famous first paragraph of this paper:

> The views of space and time which I wish to lay before you have sprung from the soil of experimental physics, and therein lies their strength. They are radical. Henceforth space by itself, and time by itself, are doomed to fade away into mere shadows, and only a kind of union of the two will preserve an independent reality.

To draw a Minkowski diagram, we take the xy-plane, call the x-axis "space" and call the y-axis "time." Since there is only one space dimension, a Minkowski diagram can be thought of as the space-time of Lineland. In Figure 78 we drew a sort of Minkowski diagram for a 2-D space. A complete Minkowski diagram for our 3-D world would, of course, take four dimensions, but it turns out that the Lineland Minkowski diagram is adequate for our purposes (Figure 82).

There is a familiar sense in which motion is relative. If two

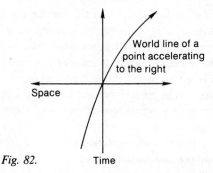

Fig. 82.

rocket ships are drifting in empty space with their engines cut off and if they pass each other heading in opposite directions, then it is impossible to say for sure if one or the other or both of them are moving. All that is certain is that they are moving relative to each other (Figure 83).

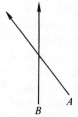

"*A* is motionless and *B* is moving rapidly to the right"

"*A* is moving slowly to the left and *B* is moving slowly to the right"

"*A* is moving rapidly to the left and *B* is motionless"

Fig. 83.

Is it *really* impossible to decide who is moving? We know from experience that no mechanical experiment will tell us if we are in a state of uniform translatory motion (that is, no acceleration and no swerving). Thus, for instance, if you're doing a steady 65 on the thruway and you toss a beer to the guy in the back seat, the beer doesn't smash into him at 65 miles per hour and kill him. Or if you want to practice yoyo tricks in the aisle while flying to the convention in Tulsa there's no need to find out your air speed and adapt your style accordingly.

But maybe there's some tricky experiment using light rays or a cyclotron or a fantastically accurate scale along with perfect clocks and rulers which would enable you to tell if you were moving or not. Einstein says no in his *Principle of Relativity:* "The laws by which the states of physical systems undergo change are not affected, whether these changes of state be referred to the one or the the other of two systems of coordinates in uniform translatory motion." Before the rabbits start coming out of the hat we need one more principle, the Constancy of the Speed of Light: whenever you measure the speed of a light ray, you're going to get the same number; it doesn't matter if you're moving toward or away from the source of light, and it doesn't matter if the source is moving toward or away from you. Given the Principle of Relativity, of course, we could delete the section "it doesn't matter if you're moving toward or away from the source of light, and" from the last sentence, since the Principle of Relativity says we can always assume we are motionless and ascribe all of the relative motion between us and the light source to the light source.

The Principle of the Constancy of the Speed of Light is hard to swallow at first. If you run forward as you throw a rock, it goes faster than it does if you throw it while standing still. So shouldn't the light coming from the headlight of a car speeding toward you be moving faster than the light coming from the headlight of a parked car? Let us temporarily entertain the notion of a luminiferous (light-carrying) aether, an invisible elastic sort of substance that fills the empty space between atoms. Light, then, is viewed as a wave in the aether, much as sound is a wave in the air and a breaker is a wave in the water. Now the speed at which sound travels through the air has nothing to do with the speed of the source. A gunshot produces a high-pressure region that is transmitted through the air at a speed that does not depend on the motion of the gun. The ripple caused by a rock thrown out into the lake moves at the same speed as the ripple caused by a rock dropped into the lake. Thus we might imagine that the speed at which a light ray approaches us need not depend on the motion of the light source.

And this is in fact the case. The various stars in the sky have a wide range of velocities relative to us, but all their light rays reach us at the same speed. This is an experimentally tested fact. So, fine, you may think, the reason the speed of light doesn't depend on the motion of the source is that light is a vibration of the aether whose rate of transmission depends solely on the aether; once a vibration is imparted to the aether, the aether doesn't care where the vibration came from—it just sends it along at the usual speed.

But what if *you* are moving relative to the aether—shouldn't that change the speed of light? If you drive a speedboat in the proper direction with the proper speed you can keep it right between the same two ocean waves; so shouldn't you be able to at least slow light down by moving away from the source through the aether? The Principle of Relativity says that your moving away from the source is no different from the source's moving away from you, and we know (by observation) that *that* doesn't change the speed of light, so it must be that motion relative to the aether *doesn't* change the speed of light rays which you observe. So the aether is even less concrete than we had imagined; indeed it isn't the kind of thing relative to which you can have a motion. It turns out that if we regard the aether as an idealized space-time instead of a mere idealized space we won't get in trouble.

What *is* the speed of light anyway? It's usually represented by the constant symbol c. It turns out that c, the speed of light, is around one billion miles per hour. For our purposes it will be convenient to pretend that c is exactly one billion miles per hour. The scale on the space and time axes of a Minkowski diagram is normally adjusted so that light rays have a slope of ± 1. Thus, if our space unit is one billion miles and our time unit is one hour, then light travels at a speed of one space unit per time unit (Figure 84).

In the relativistic world view, space-time is a sort of absolute background onto which we project our distinct conceptions of space and time. The "points" of space-time are called events. An event is a specific location in space-time. Your birth is an event in space-time; my typing the period of this sentence is an event in space-time. There is no preferred way for assigning space and time coordinates to the events of space-time, but the tracks of light rays do supply a sort of built-in structure to space time. That is, whether or not there is a light ray connecting two events is not a matter of opinion, something on which different observers could disagree. If event A is the explosion of a hydrogen bomb on the moon and event B is your noticing a flash of light on the moon, then no observer can dispute the fact that there is a light ray connecting event A to event B (Figure 85). The really significant thing about the light rays is that every observer agrees on the speed of light.

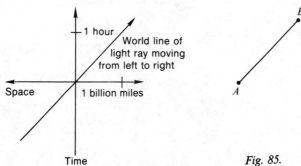

Fig. 84. Time   Fig. 85.

Let us see how this affects the definition of simultaneity. Suppose that there is a long platform (say a train) moving rapidly to the right. Let there be an observer O stationed at the middle of the platform and say that there is a small bomb at each end of the platform (Figure 86). The bombs are going to be detonated, and O is going to want to decide if they were detonated at the same time. That is, he wishes to know if the event which is the detonation of the bomb on his left and the event which is the detonation of the bomb on his right have the same time coordinate. It seems quite reasonable of O to say, "I will conclude that the two bombs were exploded simultaneously if I see the flashes of the explosions at the same instant. For since the bombs are located at equal distances from me the two flashes will reach me at the same time only if they occur at the same time." Notice, however, that if O perceives the two flashes at the same time then it must be that *we* will think that the bomb on the left exploded first. For we will reason that since he is moving away from the left flash and toward the right flash, in order for the left flash to catch up with him at the same time that he meets the right flash, it must be that the left flash got started first (Figure 87). O, of course, is free to regard himself as motionless. According to the Special Theory of Relativity his idea of simultaneity is as valid as ours. O divides space-time up into space and time in a way different from us. We see space-time as a continuum of horizontal space cross sections stacked up in the time direction. Only, if someone is moving relative to us, his time direction is different and his idea of how to slice the stack into space cross sections (a space cross section being a collection of simultaneous events) is different as well. The first difference is not so surprising; certainly his time axis can be different from ours if he is moving. Your time axis, after all, is the collection of all events whose space coordinate is zero, and if you assume you are motionless at the origin of space then your world line will *be* your time axis.

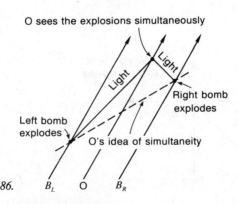

Fig. 86.

   The really surprising thing is that as your time axis changes your
space axis changes along with it. It turns out that the angle your
space axis makes with the horizontal axis is the same as the angle
your time axis makes with the vertical axis. We should keep in mind
that there is *no way* to decide which coordinatization is "right." There
is no such thing as a "right" coordinatization; any division of
space-time into space and time is equally arbitrary. Thus O's version
of Figure 87 will have *his* axes orthogonal and *ou*r axes slanting
(Figure 88).

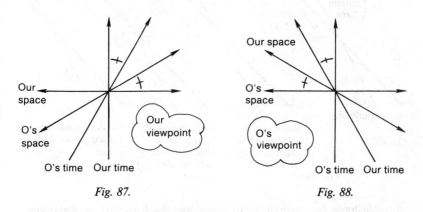

Fig. 87.   Fig. 88.

   The main point is that it is literally meaningless to claim that
distant events are or are not simultaneous. Simultaneity is not an
intrinsic property of space-time; it is only an artifact of the manner
in which we perceive, splitting 4-D space-time into a continuum of
3-D spaces arranged along a time axis.

   This notion of simultaneity is an important one. Let us discuss it
a little more. How, if you receive a light signal from some event, do
you decide when the event actually took place? The speed of light is
to be constant at a billion miles per hour for each observer, and this
supplies us with a conversion factor between space and time. That is,
if you receive a light signal from a place which you know to be one
billion miles away, then you can conclude that the signal was emitted
one hour ago. But what if you're moving away from the place where
the signal took place? No such thing, says Einstein. Once a light
signal gets going you are free to assume that the aether that carries
the signal is moving along *with* you. You don't have to account for
your motion relative to the source. If you knew how to account for it
you'd know that you were moving, contradicting the Principle of
Relativity. You already know that if the source is moving away from
you, that shouldn't make any difference, since the light wave is a

process that "forgets where it came from," and you're free to ascribe to the source any motion of yours relative to the source. So if you're moving away from the source of the light flash at about half the speed of light (from the standpoint of the person drawing this Minkowski diagram), what's going to happen? How are you going to decide at what point on your world line event X occurred (Figure 89)?

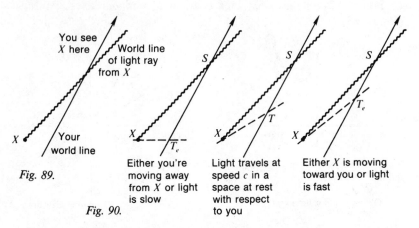

You see X here

World line of light ray from X

X

Your world line

Fig. 89.

Either you're moving away from X or light is slow

Light travels at speed c in a space at rest with respect to you

Either X is moving toward you or light is fast

Fig. 90.

We have two principles to guide us: the Principle of Relativity and the Principle of the Constancy of the Speed of Light. What you have to do is to pick some event T on your world line and say, "X occurred simultaneously with T." T might be, for instance, the event of your watch reading 12 noon. Call the event when you see the light signal from X the event S (Figure 90).

Now, T must be chosen so that the space separation between X and T is equal to c times the time separation between T and S. If you choose an earlier time T_e, then the light from X is going to take a longer time to cover a shorter distance, and you will be forced to conclude either (a) I am moving away from the space location X, or (b) the light from X is approaching me at a speed less than one billion miles per hour. Similarly, if you choose a later event T_l, then you will be forced to conclude either (a') I am moving toward the space position X, or (b') the light from X is approaching me at a speed greater than c, since here the light covers an apparently greater distance in a shorter time.

Note that conclusions (a) and (a') violate the Principle of Relativity, whereby you are always allowed to assume that light travels through an aether that is at rest with respect to you; that is, you are allowed to assume that you are at rest with respect to any given event. Conclusions (b) and (b'), on the other hand, violate the Principle of the Constancy of the Speed of Light, which states that every observer must perceive the speed of every light ray to be the same.

So now we can see how to find the event T on any straight world line which an individual traveling that world line must believe to be simultaneous with any given event X. You draw a light ray's world line from X, keeping in mind that the world lines of light rays are always at a 45° angle to the horizontal. Find the event S where the light ray's world line crosses the reference world line. Pick a point T on the reference world line so that the distance XT equals the distance $S'T$. T is the event which a person traveling the reference world line must conclude is simultaneous with X. You can actually construct the point T by taking it to be the point where the perpendicular bisector of XS crosses the given world line (Figure 91).

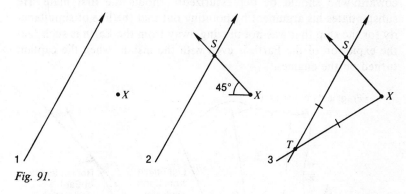

Fig. 91.

The relativity of simultaneity can lead to some paradoxical situations. Imagine that a rocket ship is floating in space, out near Pluto, and that it is staying at a fixed distance from the Earth. At some time the captain decides to move on out of the Solar System, so he turns on the ship's engines and it accelerates away from the Earth. After a while he cuts off the engines and the ship continues to coast away from the Earth, only now at a constant speed. They coast along for a while and then they decide to look through their telescope and see how good old Earth is doing. To their horror, what they see through the telescope is the destruction of the Earth by a doomsday device, a single bomb so powerful that it shatters the Earth into fragments the size of asteroids.

They realize, of course, that the destruction of the Earth is not happening as they watch; it will have taken the light from the explosion some time to get from the Earth out to the rocket. But they are interested in figuring out exactly *when* the destruction of the Earth did take place. In particular, they would like to know if it took place before or after the captain accelerated the ship away from the Earth.

The captain argues that the explosion of the Earth took place right after he turned off the engines. "I had a feeling that Earth needed us, so I cut power," he claims. He substantiates his argument by pointing out on a handy Minkowski diagram that the line of simultaneity for the ship that is coasting away from the Earth is such that the expolsion of Earth is even with the instant when he cut power.

The first mate argues that the explosion of the Earth took place right before the captain turned on the engines. "The captain knew that the Earth was about to go up in flames, so he decided he'd better head on out of the Solar System. I think the captain is a traitorous coward who should be depressurized!" shouts the first mate. He substantiates his argument by pointing out that the line of simultaneity for the ship that was not moving away from the Earth is such that the explosion of the Earth is even with the instant when the captain turned on the engines.

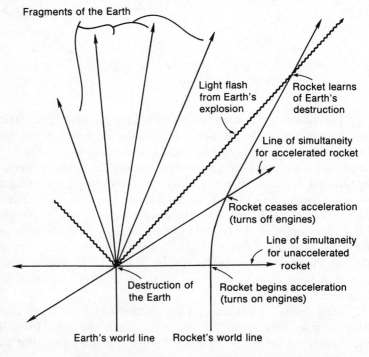

Fig. 92.   Did the Earth blow up before or after the ship accelerated?

Who is right? Did the Earth blow up before the acceleration, after the acceleration, at both times, or at neither time (Figure 92)? Actually, there is no real answer! What does it mean when you ask a question about the world and there is no real answer? It means that you're asking the wrong kind of question. In this case the moral is that it is really meaningless to speak of two distant events as being simultaneous.

There is worse to come. We will soon see that objects do not have length in any absolute sense. Imagine two segments in Lineland that move past each other at a high rate of speed. One segment moves to the right at half the speed of light, and one segment moves to the left at half the speed of light. Riding on the midpoint of the segment moving to the right is a point called R, and riding on the midpoint of the segment moving to the left is a point called L. Before they started moving, both segments were the same length, and to us they still appear to be the same length. However, R will say that L's segment is shorter than his, and L will say that R's segment is shorter than his. How is this possible? Let's look at the Minkowski diagram (Figure 93). We identify ourselves with a fixed point 0 of Lineland.

We have drawn the world lines of 0, R and L and we have drawn the world lines of the ends of the segments on which R and L are riding. Notice that there is an event in space-time where 0, R and L are at the same place at the same time. We have drawn the lines of simultaneity for the three observers which pass through this event. 0's line of simultaneity is horizontal, R's slants up and L's slants down. It is easy to draw in the lines of simultaneity because the angle between any observer X's space axis and 0's space axis must always equal the angle between X's time axis and 0's time axis. (This is so that the speed of light will appear constant, i.e., so that a line going out from the origin with a slope of 1 in one system will have a slope of 1 in any other system.)

We are interested here in the event A and the event B. The event A is when the tip of L's segment crosses the end of R's segment; the event B is when (and where) the tip of R's segment crosses the end of L's segment. According to 0, the events A and B are simultaneous. Hence 0 concludes that the two segments have the same length, since there is an instant of time when they overlap each other exactly.

According to R, event A occurs after event B. For, at the instant when R meets L, event B lies below R's line of simultaneity (i.e., in R's past) and event A lies above R's line of simultaneity (i.e., in R's future). So R will say, "First the tip of my segment crosses the tail of L's segment, and then some time later the tail of my segment crosses the tip of L's segment." From this R will conclude that his segment is longer. This follows if you think about it a little. Say, for instance,

The relativity of length

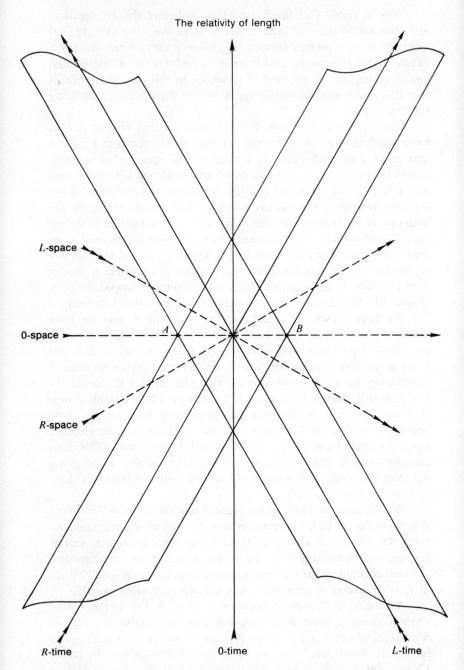

L-space

0-space

R-space

A

B

R-time

0-time

L-time

Fig. 93.

that you were driving a Cadillac to the east and someone else was driving a VW to the west (Figure 94). When your paths crossed, your hood ornament would be even with the VW's rear bumper first (at this time, the VW's hood ornament would just be even with your rear door) and then after a while your rear bumper would be even with the VW's hood ornament. You could conclude from this that you had a bigger car.

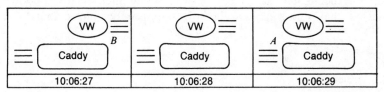

Fig. 94.

Actually there's an easier way to look at the Minkowski diagram and see that R will think L's segment is shorter than his. Just look at the line labeled "R-space." This is R's line of simultaneity for the instant when his path crosses L's path. If you just think of this as R's space axis, then you can see that on this space axis L's segment is shorter than R's segment.

The same kind of argument shows that L will think R's segment is shorter than his. L will say that event A happens before event B, so he can conclude that his segment is longer. Or, just by looking at L's space axis, you can see that R's segment is shorter relative to this notion of space.

Actually, for this argument it was not really necessary to have both R and L moving. Taking R's standpoint, we can see that R is free to imagine that he is motionless and that it is just L that is moving. So the upshot of this argument is that *moving objects appear to be contracted in their direction of motion.* R thinks he is motionless and L is moving, so he sees L as contracted; L thinks he is motionless and R is moving, so he sees R as contracted. 0 thinks R and L are moving in opposite directions with equal speeds, so he sees them both contracted by equal amounts. Note that we have not yet figured out how to indicate how long R's segment and L's segment would look to 0 if they stopped moving. Before we can do that, we will need the idea of the *interval* between two space-time events.

But first let's examine an apparently paradoxical consequence of the relativity of length, a paradox entitled the Pole and Barn Paradox. Imagine a barn 10 meters long and a man running toward it carrying a pole 20 meters long (Figure 95). The rear wall of the barn is paper, so he can run through it without getting killed. The plan is to let him run into the barn and then to slam the door as soon as the trailing end of the pole gets inside the barn. Now this guy is a really fast runner. In fact he is running at about three-fourths the speed of light. As it turns out, if he runs this fast, then the pole he is carrying will appear 10 meters long to the farmer in the barn. On the other hand, given the Principle of Relativity, the runner is going to see the *barn* as being half as long as it was before their relative motion started; that is, he's going to think the barn is only 5 meters long.

Be plenty o' time t' close thet door ...

The farmer sees it like this

Hope that's thin paper...

The runner sees it like *this*

Fig. 95.

Now, it would seem that we could decide in an absolute sense who was right, the farmer or the runner. For once the pole gets all the way past the barn door and the farmer slams the door, then either the runner and his pole will be entirely inside the barn, or the runner will already have burst through the rear wall of the barn. Right? Wrong. Whether the runner bursts through the rear wall before or after the farmer gets that door closed involves a judgment of which events are simultaneous! And simultaneity of events at different places is a relative concept!

"Wal, first I swanged the door to, 'n then I heerd him busting out through the barn wall," the farmer says. The runner says, "When the pole went through the rear wall I glanced back and saw the pole still sticking way out of the barn door. He didn't get that door closed till I had crashed through the wall myself. And look, I thought you said that wall was going to be *paper*!" The runner will feel that the farmer thought he fitted inside the barn only because he took the sense impression of the pole being about to hit the concrete-block rear wall and pretended it was simultaneous with the sense impression of closing the door. The farmer will feel that the runner thought his pole didn't fit inside the barn because he pigheadedly pretended his crashing through the wall took place before the door closed.

Who's *really* right, the farmer or the runner? As with the paradox of the Earth's explosion, there is no real answer to this. The problem is that all that *really* exists is world lines in space-time. There is no built-in division of space-time into a time component and a space component. Different observers will accomplish this division in different ways.

As we have seen, given two distinct events *A* and *B*, there is no absolute way of deciding if *A* and *B* are simultaneous and there is no absolute way of deciding what is the distance *between A* and *B* (relativity of length). It turns out that there is also no way of finding an absolute time span between the events *A* and *B*, either, but we'll leave that for later.

What we would like to do now is see if there is *any* sort of relation between events *A* and *B* that does not depend on the observer.

We already know of two such relations: (i) if *A* and *B* occur at the same place and time, then every observer will agree on this fact, and (ii) if event *B* is the reception of a light signal whose emission was event *A*, then everyone will agree on this fact. Given these two facts and the Principle of the Constancy of the Speed of Light, it is possible to prove mathematically that the *interval* between events *A* and *B* will be the same for every observer. We now explain what "interval" is.

Take one observer's frame of reference. Say that he assigns coordinates (x, t) to event *A* and coordinates (x', t') to event *B*. Then the interval between *A* and *B* is said to be the number r such that

$$r^2 = c^2(t' - t)^2 - (x' - x)^2.$$

Here c is the speed of light (approximately one billion miles per hour), so we can see that the interval r will be in units of distance (miles). As we mentioned before, one frequently chooses the units in relativity theory to be such that the speed of light c is *one* distance unit per time unit. Let us assume this has been done. Another simplification occurs if we write Δt for $t' - t$ and Δx for $x' - x$. "Δ" is pronounced "the change in" (or "delta"). Now our definition of interval has the form $r^2 = \Delta t^2 - \Delta x^2$. Writing ΔI, "change in interval," for r, we have $\Delta I^2 = \Delta t^2 - \Delta x^2$, or "change in interval squared equals change in time squared minus change in space squared."

We can see that interval in the xt-plane is quite different from distance in the xy-plane. For interval we have $\Delta I^2 = \Delta t^2 - \Delta x^2$, whereas for distance we have (writing s for distance) $\Delta s^2 = \Delta y^2 + \Delta x^2$. This last equation is just the well-known Pythagorean theorem!

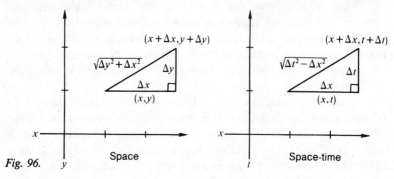

Fig. 96.

Space   Space-time

Notice that the interval between two events will be zero if $\Delta x = \pm \Delta t$. Under what circumstances will the space separation between A and B equal the time separation between A and B? Exactly, when there is a light ray connecting A and B, since light travels at the speed of *one* space unit per time unit. In the $x't'$ coordinate system, their idea of a meter or of an hour may differ from what the guys in the xt coordinate system think, but the difference in space unit and time unit between the two systems will always be coordinated so that the speed of light is one. That is, the speed you get by comparing event A and event B will be either $\Delta x/\Delta t$ or $\Delta x'/\Delta t'$. But either way it's got to come out to c, which is *one* in this discussion. Hence you have to have $\Delta x = \Delta t$ and $\Delta x' = \Delta t'$, and thus $\Delta I = 0$ and $\Delta I' = 0$.

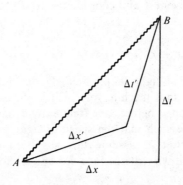

Fig. 97.

We can already see that interval in space-time is quite different from distance in space. If there is zero distance between two points we know that they are identical, but if there is zero interval between two events, it does *not* follow that the events are identical. If two events have zero interval between them it means only that there is (or could be) a light ray connecting them. For instance, if there was an explosion a billion miles away from us an hour ago, a light ray from the explosion could reach us right now, indicating that the event of the explosion and the event of our standing here right now have zero interval between them. Actually, for *A* and *B* to have zero interval between them it is not necessary that a light ray actually be sent from *A* to *B*; it is only necessary that this be possible. In other words, it is only necessary that $\Delta x = \pm \Delta t$.

Consider the event that is your existence at the instant you read this sentence. You can imagine youself as being the origin of a 4-D space-time system. Since we are living in 3-D space instead of a 1-D space like Lineland, we must speak of your "space of simultaneity" instead of your "line of simultaneity." Your space of simultaneity is all the events in space-time that you believe to be happening at this instant. It includes, for example, a man lighting a cigarette in the rain somewhere in South Wales, a momentary increase of temperature at the North Pole of the Sun, and the death of a cell in your best friend's body.

Your time axis is a line in 4-D space time that includes your world line, i.e., includes every event of your life, past and future. [We remark here that there is an inaccuracy in the assertion that your world line can be taken to be part of the time axis of a space-time coordinatization satisfying the Special Theory of Relativity. The problem is that your world line is not "straight." For instance, the planet you live on is rotating; for instance, you're always jumping up and changing your velocity by walking around. But if you were floating in empty space and not somehow accelerating and decelerating, then this discussion would be accurate.]

Now, what I wish to talk about here is your *light cone*. Your light cone is the collection of all the events whose interval from you is zero. Your light cone is the collection of all the events *A* such that either (i) a light flash that took place at *A* would be seen by you right here and now, or (ii) a light flash that took place right here and now (if your head exploded, for instance) would be visible at the very time and place corresponding to *A*. Your light cone has two halves, the back light cone (events satisfying [i]) and the forward light cone (events satisfying [ii]).

To get to any event on your forward light cone you'd have to travel at the speed of light. As far as we know, material objects cannot ever go as fast as light. So any event that you're going to be able to make it to is going to lie inside your forward light cone. These events are collectively known as your Future. The events inside your back light cone are called your Past. If the event consisting of some weird creature's getting in a space ship and blasting off lies in your Past, then it is possible that he is going to arrive, right Here & Now. If the event consisting of his blasting off does not lie in your Past, then there's no way he could turn up Here & Now without going faster than light.

What about the events which don't lie on or inside your back light cone and don't lie inside or on your forward light cone? This collection of events is called Elsewhere (Figure 98). It is soothing to think about the events that are in your Elsewhere. There is no way such an event can affect you right now, and there is no way that anything you do can be affecting such an event. "Oh my gosh, what if the Russians just launched a nuclear attack against us?" "Cool it, man. That's Elsewhere." You see, if the Russians were indeed pushing that button right this instant, it still wouldn't affect your Here & Now. Of course, in about ten seconds the event of their pushing the button would be in your Past, but for now it's still Elsewhere.

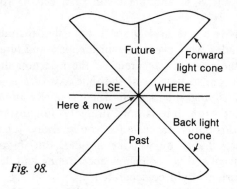

Fig. 98.

It is a rather striking fact that your whole space of simultaneity is in your Elsewhere. This 3-D universe consisting of the events that you say are occurring at this instant, there's no way you can change anything in it, and no way anything in it can affect you Here & Now. By the time you see and recognize anything it's in your Past. If you throw a rock, it lands in the Future.

Going back to the Lineland Minkowski diagram, there is a simple way to decide if an event (x,t) is Elsewhere with respect to an observer at the origin $(0, 0)$. The interval between $(0, 0)$ and (x, t) is the square root of $t^2 - x^2$. If $|t|$ is greater than $|x|$, then $t^2 - x^2$ is positive, and $I = \sqrt{t^2 - x^2}$ is a real number. If $|t|$ is less than $|x|$, then $t^2 - x^2$ is negative, and $I = \sqrt{t^2 - x^2}$ is an imaginary number. If $|t|$ equals $|x|$, then $t^2 - x^2$ is zero, and I is zero as well.

If $|t|$ is greater than $|x|$, then a trip between $(0, 0)$ and (x, t) involves traveling *slower* than light. If $|t|$ is less than $|x|$, then a trip between $(0, 0)$ and (x, t) involves traveling *faster* than light. This is true since speed is distance divided by time, i.e., $|x/t|$ . . . and in this discussion, the speed of light is 1.

If the interval between two points is real, the points are said to have *time-like* separation. If the interval between two points is imaginary, the points are said to have *space-like* separation. If the interval between two points is zero, the points are said to have *light-like* separation.

The remarkable thing about the interval is that it is the same no matter who measures it. The runner and the farmer disagreed on the space separation between the door of the barn and the rear of the barn, and they disagreed on the time separation between the tip of the pole breaking the rear wall and the end entering the front door. They disagreed on both the space and time separation between these two events. However, they would find the same *interval* of space-time separation between these two events.

Because the interval is the same for everyone who measures it, if we believe that the interval between two events A and B is time-like, then so will every other observer. If we believe that the interval between two events A and B is space-like, then so will every other observer. That is, if we think it would be possible for a space ship to have its blast-off be event A and its landing be event B, then every other observer would think so too. In Figure 99 the interval is the square root of 7.

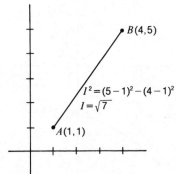

Fig. 99.

If the separation between two events A and B is *time-like*, we can meaningfully ask if B lies in A's Past or Future. If the separation between A and B is space-like, then we cannot assign any absolute order in time to the events A and B. If two events have space-like separation, some people will say they are simultaneous, some people will say A happens first, and some people will say B happens first. This is as opposed to the case where the separation is time-like and everyone agrees, say, that B lies in A's Future.

If you turn back to Figure 93, you can see two events A and B with a space-like separation, and you can see that R, 0 and L will have all the possible opinions about which event happens first.

In the xy-plane, the set of all points whose distance from the origin is 1 constitutes a circle, the unit circle. In the xt-plane, what does the set of all points whose interval from the origin is 1 look like?

If the interval between $(0, 0)$ and (x, t) is 1, then we must have $t^2 - x^2 = 1$. What is the graph of this equation in the xt-plane? The unit hyperbola! You can see that from the fact that we must always have $|t|$ bigger than $|x|$, but that for large values, $|t|$ and $|x|$ are approximately equal; so the graph is asymptotic to the lines $x = t$ an $x = -t$.

If we stick to the idea that one space unit is one billion miles and one time unit is one hour, then we can see that each point on the upper half of the unit hyperbola has a time-like separation of 1 (hour or billion miles, you can put it either way, if you understand that a billion miles of time is an hour, the length of time it takes light to travel one billion miles) from 0. How long would it seem to take to go from 0 to A along the line indicated in Figure 100?

Relative to the coordinate frame we have drawn in, it looks as if

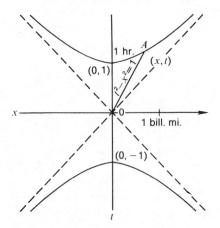

Fig. 100.

the t-coordinate of A is about 1.2. But what about the coordinate frame of someone whose time axis lies along the line segment $0A$? For him, there will be no change in space when he goes from 0 to A (for instance, he might be sitting in a space ship so large that he believes it to be motionless; he would say he "got" from one place to another because the outside world was moving). That is, if A has coordinates (x', t') in his coordinate system, then we know that $x' = 0$. Now the interval between 0 and A is to be measured as 1 by every observer. So we have $1 = t'^2 - x'^2$, or $1 = t'^2$, or $t' = 1$. In other words, the moving observer assigns space-time coordinates $(0, 1)$ to A. In other words, the moving observer thinks it takes him only one hour to get from 0 to A and we thought it took him 1.2 hours!

In fact, you can travel to any point in the universe in one hour! Say, for instance, you'd like to get to some star that is two billion miles away, and you'd like to be there in an hour. At first it seems that this is not possible, since *light* can only go *one* billion miles in an hour, and you don't expect to be able to go faster than light.

But once you take a look at the Minkowski diagram of the situation (Figure 101), you see that if you could just take off with a speed close enough to the speed of light you could have your arrival be the event A whose space coordinate is 2 and which lies on the unit hyperbola. A is a time-like interval of 1 from 0. So if you were traveling along the segment $0A$, you'd think it was your time axis (in a state of uniform motion, one *always* identifies the time axis with one's world line). So you'd assign coordinates $(0, t')$ to the point A. Now, since A lies on the unit hyperbola, we *know* that the interval between 0 and A appears to be *one* to *any* observer. So the moving observer must have the interval $0A = t'^2 - x'^2$ equal to 1. But he has $x' = 0$, so it must be that $t' = 1$. "It only took me an hour!"

How fast, precisely, *would* you have to go to get two space units

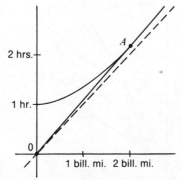

Fig. 101.

away from earth in one time unit (i.e., two billion miles in one hour, or two light-years in one year)? Of course, *you* could pretend that you were motionless, but how fast would you be going relative to the *Earth's* reference frame? The people on Earth would not think that it just took you an hour to that star. In their reference frame, the event A has a t-coordinate slightly greater than 2, as opposed to your reference frame, where the event A has a t'-coordinate of 1.

If we knew what the t-coordinate of A was, we could figure out how fast the Earthlings thought you flew. For we already know that the x-coordinate of A is 2, and your speed relative to Earth will thus be $v = x/t = 2/t$, where we are calling the t-coordinate just t. How can you figure out what t is? You know that the interval between 0 and A is one. So you know that $1 = t^2 - 2^2$. Hence $t = \sqrt{5} \approx \frac{11}{5}$ and $v = 2/t \approx \frac{10}{11}$ of the speed of light.

Consider the reference frames of two observers. Let the frame of reference of the observer with whom we identify be the xt-system, and let the frame of reference of the moving observer be the $x't'$-system. We learned in the discussion of the Relativity of Simultaneity that if the t'-axis is different from the t-axis, then the x'-axis is different from the x-axis. In fact, we learned that the angle between the t' and t-axes always equals the angle between the x' and x-axes.

It turns out that there is another difference between the xt and $x't'$-systems. This is that the unit marks on the x' and t'-axes are farther from the origin than in the xt-system.

As we have just been discussing, the unit time mark on the t'-axis will be where that axis crosses the unit hyperbola. Given this we can see where to locate the unit space mark on the x'-axis, since the time unit and space unit have equal size. When we have drawn everything in as in Figure 102, we can see that the speed of light will be 1 in the $x't'$-system as well as in the xt-system.

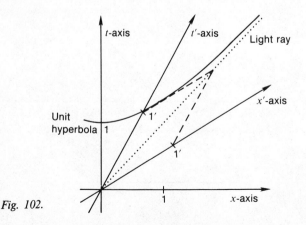

Fig. 102.

PROBLEMS ON CHAPTER 5

(1) In this problem you will work out an argument for the relativity of simultaneity slightly different from the one I already gave. The situation is as follows. A rigid platform is moving to the right at, say, half the speed of light. On the left end stands Mr. Lee, and on the right end stands Mr. Rye (Figure 103). Mr. Lee sends a flash of light down the platform toward Mr. Rye. Mr. Rye holds a mirror that bounces the light flash back toward Mr. Lee. Mr. Lee receives the return signal. Call these events A, B and C, respectively. Mr. Lee notes the times of events A and C on his world line. After a little thought, he figures out where the event X on his world line that is simultaneous with B is located. Where does he put X, and why?

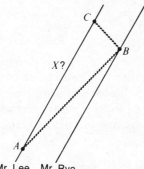

Fig. 103.   Mr. Lee   Mr. Rye

(2) People who do not believe in the static space-time view outlined in Chapter 4 like to claim that time is really moving, that "now" exists, but that the future does not in any sense exist yet. Evaluate this claim in the light of the following quote:

> The existence of an objective lapse of time, however, means (or at least, is equivalent to the fact) that reality consists of an infinity of layers of "now" which come into existence successively. But, if simultaneity is something relative in the sense just explained, reality cannot be split up into such layers in an objectively determined way. Each observer has his own set of "nows," and none of these various systems of layers can claim the prerogative of representing the objective lapse of time." (K. Godel, "A Remark about the Relationship between Relativity Theory and Idealistic Philosophy," in the Schilpp anthology, p. 558; see Bibliography)

(3) In this problem you will see why a moving person's clock seems to go slower than that of a stationary observer. Consider two people, R and L, moving in opposite directions at what appear to us to be equal speeds. Say that they pass each other at event 0, that event A is

when L's watch shows that an hour has elapsed since 0, and that event B is when R's watch shows that an hour has passed since 0. For reasons of symmetry, *we* will percieve A and B as simultaneous (as indicated in Figure 104). R, however, will say that A is simultaneous with A', and L will say that B is simultaneous with B'. Why? What will R and L say about the other's idea of an hour, and why?

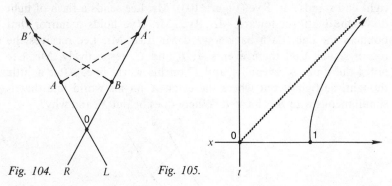

Fig. 104.   R   L   Fig. 105.   t

(4) How could you cause your world line to be the section of the hyperbola $x^2 - t^2 = 1$ indicated in Figure 105? If this was your world line, could a light signal from event 0 ever reach you? How far is 0 from you relative to any one of your instantaneous frames of reference?

(5) Suppose that you were in a *very* powerful rocketship which accelerated even faster than the one in the last problem. Say that you accelerated away from the Earth so that the first billion miles took you an hour of your time, the second billion miles took you a half hour of your time, the third billion miles took you a quarter hour of your time, and so on. In general it would take you $1/2^n$ hours to cover the $n + 1$st billion miles. Where would you be after two hours?

(6) Say that you are on a space station moving away from Earth at $1/2$ the speed of light relative to Earth (world line of slope 2 in the Earth's space-time diagram). You then get into a small ship and blast off, moving away from Earth and the space station at $1/2$ the speed of light relative to the space station (world line of slope 2 in the station's space-time diagram). Are you then moving away from Earth at the speed of light? Combine Figure 106 and 107, and make an estimate of how fast you will be moving relative to Earth.

(7) Say that you travel from event A to event B. If you carry a clock with you on your travels, it turns out that, since you are free at any instant to pretend that you are motionless ($dx = 0$), your clock will

measure the interval you have gone. Which of the three indicated paths between *A* and *B* in Figure 108 will have the *longest* interval? Do you think time-like geodesics in space-time will maximize or minimize the interval?

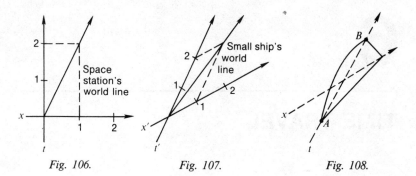

Fig. 106.   Fig. 107.   Fig. 108.

6

TIME TRAVEL

The Special Theory of Relativity implies that it is impossible for any material object to travel as fast as light, and that it is impossible for any type of signal to travel faster than light.

That no material object can travel faster than light is a fact that has been experimentally tested. Given an electron in a cyclotron, one can pile as much energy onto it as one wishes and it never reaches the speed of light. One reason for this is that as an object moves faster its mass increases, so that the faster it goes, the harder it is to make it go any faster.

Does this mean that we can never travel at the speed of light? Not necessarily. It would perhaps be possible (this is science-fiction) to break a person down into a complicated electrical wave-form and transmit this wave by radio (radio waves travel at the same speed as light) to a deprocessing station where the person would be reconstituted out of the information in the radio wave.

What would it feel like to travel at the speed of light? Say you went from here to the other side of the galaxy at the speed of light; how long would it seem to take? It would seem to the people at the sending and receiving stations that it took the signal some hundred thousand years to cross the galaxy. But to you the trip would seem instantaneous! You'd step into the dematerialization booth in one door and walk out the other side of the booth without even slowing down. Only, when you stepped out the other side it would be 100,000 years later and on the other side of the galaxy. If you suddenly got homesick and walked back through the booth the other way, you'd be back on Earth, only it would be 200,000 years after you started. And to you these 200,000 years would have seemed to consist just of walking back and forth through a booth. Going back and forth like this five times would put you a million years in the future, and so on.

The reason that travel at the speed of light seems to take no time at all is that when you reach the speed of light, your world line lies in your space of simultaneity. That is, for someone going at the speed of light, every event on his world line happens at the same time—and in the same place! We can see this by looking at the three Minkowski diagrams in Figure 109; as the time line slants over, the space line slants up, and finally meets it when you go at the speed of light.

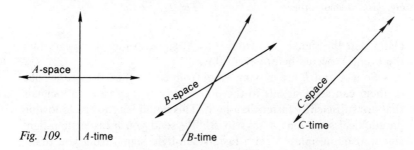

A-space

B-space

C-space

C-time

Fig. 109.   *A*-time   *B*-time

So if you can travel at the speed of light, you can get to any event on your future light cone in no time at all. By bouncing back and forth you can also get to any event inside the future light cone (like right here a million years from now) in no time at all.

You can't get back, though. Why not? Is there any reason why we shouldn't somehow be able to travel into the past? Maybe not, but there are certain difficulties involved. Suppose you devise some method of traveling into the past. You travel back in time an hour and see your earlier self getting the time machine ready. With an ironical smile you shoot your earlier self in the back of the head. What happens then? Since your earlier self is dead, you cannot have entered the time machine to come back and kill your earlier self. So your earlier self cannot be dead. But if your earlier self is not dead, then you *were* able to come back and shoot it. Your earlier self dies if and only if it doesn't die. A paradoxical situation indeed.

It is this type of paradox that seems to preclude the possibility of sending signals faster than the speed of light.

Consider the Minkowski diagram in Figure 110. The dotted line $0X$ represents a signal that A sends from 0 to X faster than the speed of light. (*B* is impressed by the claim that the signal goes *from* 0 *to* X, since to him it appears that event X occurs before event 0.)

Conversely, the dotted line $0Y$ represents a signal that B sends from 0 to Y faster than the speed of light. (*A* is impressed by the

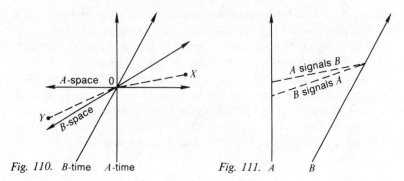

Fig. 110. B-time A-time   Fig. 111. A   B

claim that the signal goes from 0 to Y, since to him it appears that the event Y occurs before event 0.)

So if A and B are moving away from each other, then either one of them can send signals to the other's past if they can send signals that are sufficiently faster than light. This could lead to the following paradoxical situation: A says to B, "I'll send you a faster-than-light signal at noon unless I get a faster-than-light signal from you first," and B says to A, "I'll send you a faster-than-light signal whenever I get a signal from you." Now, if A sends a signal at noon, B will send back a signal that reaches A before noon, so A won't send a signal at noon. If A doesn't send a signal at noon, then B won't get a signal and won't send one back, so A will send a signal at noon. In other words, A sends a signal at noon if and only if A does not send a signal at noon (Figure 111). This seems to be impossible.

Actually, it has been proposed in recent years that there *are* in fact things that travel faster than light. These things are called tachyons ("tachy" means "fast"). But didn't we just show that you couldn't have things going faster than light?

It would actually be all right to have tachyons around if we couldn't detect them, so that you couldn't use them to send signals. Physicists, however, tend to be pretty stingy with objects that are impossible to detect. They say that if there is no possible way to detect something, then it is meaningless to talk about the thing's existence. Whether or not one agrees with them is essentially a philosophical issue. In any case the modern consensus seems to be that detectable tachyons do not exist.

There are two ways other than traveling faster than light by which one might go into one's past.

The first way is that time might be circular. That is, the universe would have no beginning or end. Such a universe with a straight infinite space dimension would look like Figure 112.

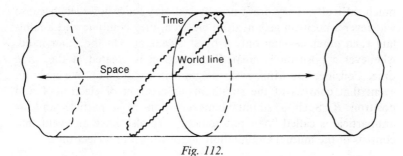

Fig. 112.

As we know, if you travel at the speed of light, your trip takes no time at all (for you). Another way of putting this is that if you travel at a speed close enough to the speed of light, you can make your trip take as short a time as you desire.

So if the length of time (the "temporal circumference of the universe") was a trillion years, you could travel away from the Earth for 1/4 trillion years, travel back toward and on past it for 1/2 trillion years, then turn around and take another 1/4 trillion years to get back. Now, if you traveled at the speed of light, this trip would take you no time at all, but you would have gone "forward" a trillion years through time. So you would get back to Earth just when you left. If you traveled just a little less further forward in time, say a trillion years minus 2000 years, you'd be able to prevent the Crucifixion!

This technique of time travel would be possible provided the universe had the right kind of structure (circular time). Kurt Gödel proposed something like this idea in his paper included in the Schilpp book of essays on Einstein's work (see Bibliography). In order to avoid the paradoxes of time travel, he asserts that it would never actually be possible to make a trillion-light-year journey of the type we have described—for practical reasons. First of all, it would take a rocket the size of a galaxy to have enough fuel to make the trip; second, you'd be thrown off course by the gravitational attraction of the various stars and galaxies you passed, and you'd never find your way back to the Earth. We'll discuss other possible structures for the space-time of the universe in the next chapter.

A second way in which objects might travel backwards in time is for them to be composed of antimatter. As far as is known, every type of particle has a corresponding antiparticle. An antielectron is called a positron. A positron has the same mass as the electron, but its charge is the exact opposite of the electron's. Positrons can be created in the laboratory (using particle accelerators) without too much difficulty. They don't usually last very long, however, because whenever a positron gets near an electron, they combine and annihilate each other, leaving only a burst of energy. On the other hand, whenever a positron is created an electron is created at the same time. Consider the Minkowski diagram in Figure 113. We have an event that consists of the simultaneous creation of electron *A* and positron *B*. Such a simultaneous creation of a particle and its antiparticle is called "pair production." We also have an event that consists of the mutual annihilation of positron *B* and electron *C*.

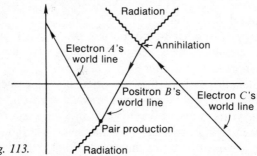

Fig. 113.

The physicist Richard Feynman has suggested that instead of viewing this diagram as representing two electrons and a positron, we can view it as representing a single particle that travels forward in time as electron *C*, travels backward in time as positron *B* and then moves forward in time again as electron *A* (Figure 114).

A really fascinating aspect of the Feynman approach is that there could be only one electron in the whole universe!

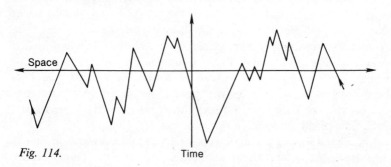

Fig. 114.

The electron would have a complex world line, sometimes going forward in time, sometimes backward. When it was going forward in time it would be an electron, when it was going backward in time it would be a positron. But there would "really" be only one electron. This provides a nice simple explanation of why all electrons have the same charge!

A weak point of this theory is that there appear to be many more electrons than positrons in the universe. Your body, for instance, is full of electrons but it is doubtful that there are ever more than a few positrons in it at any given time. A way out of this difficulty is to claim that there are regions of the universe where the balance is reversed. That is, some of the galaxies we see in the night sky may be made almost exclusively of antimatter. Their antiatoms would consist of positrons orbiting around an antinucleus of antineutrons and antiprotons.

What would happen if you traveled to such an antigalaxy and landed on an antiplanet? You and your ship would combine with a chunk of the planet in annihilation, producing a great burst of energy. In Feynman's terms, all the particles in your ship and body would, with a great burst of energy, start moving backward in time. Would you then have the experience of moving backward in time? Probably not; turning that sharp corner would scramble you up too much.

If we could get close to an antigalaxy and watch the people on an anti-Earth, what would we see? There is no real consensus on this, but it might be that we would see people living backward in time. An antiperson's life would go something like this.

Everyone was crying. The tears welled out of their handkerchiefs and ran into their eyes. They walked backward up to the grave, where the casket was slowly brought up. The body was taken home and laid in bed. As soon as the priest left, the body began breathing. The antiman and his antiwife lived backward together for 30 years. On their wedding day they took dirty clothes out of the hamper, went to church together. After the service they saw each other a few times, but then knew no longer of the other's existence. The antiman went to college, where he unlearned a great deal. He had a good knowledge of calculus, but after completing the course he knew nothing at all about it. Homework consisted of receiving papers from the teacher, which he erased, using the point of his pencil. He met his parents, and eventually stopped walking. He lay in a crib, where his mother would bring empty bottles of milk, which he would fill from his mouth. His body took in excrement from the dirty diapers which his mother put on him. Things got calmer and hazier until on one joyful day he went to the hospital with his mother, where the doctor helped him to get inside her womb. There he slowly dissolved and in nine months he ceased to exist.

Depressing, huh? Actually, of course, the anti-Earthlings would feel as if they were living lives like ours, and if they saw us, they'd think *we* were leading the kind of antilives just described. Who's *really* right? You should know better than to ask questions like that by now!

PROBLEMS ON CHAPTER 6

(1) While on the moon, one of our astronauts tried to send a telepathic message to a partner on Earth to see if such a message would travel instantaneously. What is wrong with the notion of instantaneous communication?

(2) Say that you and I were floating in space and holding on to either end of a thousand-mile-long rod. Why would you not be able to communicate with me instantaneously by jiggling your end of the rod?

(3) Show that if it were possible to build a time machine to travel into the past, then it would not be necessary for anyone to actually *invent* (as opposed to *build a copy of*) this machine.

(4) Science-fiction writers sometimes avoid the paradoxes of time travel by assuming that there are parallel universes, and that when you go back in time you actually leave the space-time of your original universe and enter the past of some parallel universe. See how this idea can be used to resolve the paradox of the time traveler who kills his "past self."

(5) Say that time is circular. Build an indestructible radio beacon and set it afloat in space near the Earth. Assume that this beacon will endure and continue broadcasting forever. Once you set one such beacon afloat, how many more should you be able to detect? What if you decide to set your beacon afloat if and only if you detect no beacons out there before the launch?

(6) Gödel's model is actually rather different from the circular-time model (invented by Reichenbach) which I discussed. Gödel's universe has a sort of "rotation" that makes possible all of the world lines indicated in the Flatland Minkowski diagram in Figure 115. D, for instance, would appear to A to be moving backward in time. What would A say about C's life?

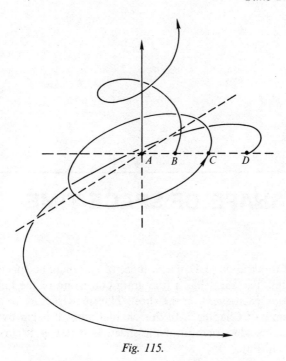

Fig. 115.

7

THE SHAPE OF SPACE-TIME

If we just think about 1-D space, then we can imagine space-time as a vast sheet. The sheet has a fine grain consisting of the light cones generated by each event on the sheet. The fine structure of the sheet was the topic of Chapter 5. In this chapter we will begin by discussing the large-scale structure of the sheet. Is it flat, is it curved, is it finite, is it infinite?

The structure of space-time, taken as a whole, is the subject matter of the science called cosmology. Since you are asking about *all* space and *all* time in cosmology, you are interested in the entire universe, everywhere and everywhen, viewed as a static geometrical object.

Sir Isaac Newton proposed the simplest view of the universe: infinite flat space and infinite time. In terms of Lineland, the Newtonian universe is simply the infinite xt-plane. If you wished to modify the Newtonian universe by claiming that space came into existence at some specific past time, you'd have the upper half of the xt-plane (Figure 116).

Around 1917 Albert Einstein proposed that space-time should be

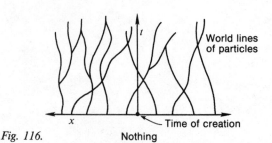

Fig. 116.

cylindrical. That is, he suggested that our space should be spherical (i.e., 3-space should be the surface of a hypersphere) and that time should be straight (Figure 117).

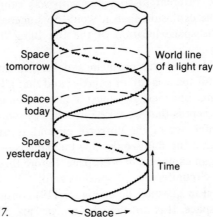

Space tomorrow

World line of a light ray

Space today

Space yesterday

Time

Fig. 117. ← Space →

It was soon discovered, however, that our space is *expanding*. That is, any galaxy we can see is moving away from us, and the further away they are, the faster they are moving away from us.

If you have a flat infinite space, you can have this type of expansion without too much difficulty. Just set down a big chunk of primordial matter (this chunk is sometimes called ylem) at some place in space-time and let it explode. An observer on any one of the pieces will see the other pieces steadily receding from him. The faster pieces will be pulling away from him, and he'll be pulling away from the slower pieces (Figure 118).

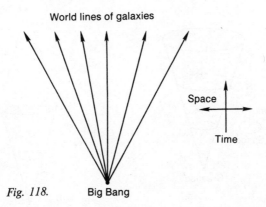

World lines of galaxies

Space

Time

Fig. 118. Big Bang

One thing wrong with this model is that it violates the Cosmological Principle. The Cosmological Principle says that things should look more or less the same no matter where you are in space. But if the universe was the result of a Big Bang in ordinary 3-D space, then it would look different to someone who was riding on one of the pieces from the explosion than it would to someone who was so far away from the space location of the Big Bang that no piece had reached him yet.

You could avoid this problem and keep ordinary space if you assumed that all space is full of galaxies and that all space has always been expanding. But there is a nicer solution: conical space-time.

The idea here is that we take Einstein's cylindrical universe, but let the circumference of the universe expand as time goes on. You again start with a Big Bang, only now the bang is not an explosion *in* 3-D space but an explosion *of* 3-D space. Before the Big Bang there *is* no space; the circumference of space is zero!

The model in question here (Figure 119) consists of an expanding spherical space. It is not that the galaxies are moving apart from each other in a flat space; it is rather that 3-D space would be the hypersurface of an expanding hypersphere.

Here, again, there is an event that can be labeled the Big Bang or the creation of the universe. It is not really meaningful, however, to ask where it was in our universe that the Big Bang took place, since when it took place there was only one point in space. That is, the Big Bang took place everywhere.

Will our universe continue expanding indefinitely? This is a controversial point. Some cosmologists believe that eventually the gravitational attraction between the galaxies will slow the expansion down and even reverse it, so that the whole universe will collapse back to a point at some future time (Figure 120).

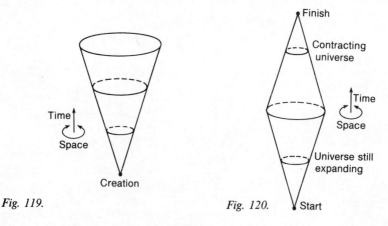

Fig. 119.

Fig. 120.

Other cosmologists maintain that the universe will continue expanding forever. There is a good chance that the answer to the dispute will be experimentally decided in our lifetime.

If you assume that the universe *does* contract back to a point at some future time, you are left with two questions: what happens after the end of the universe, and what happened before the beginning of the universe?

One viewpoint is that it is meaningless to talk about events before the beginning of time or after the end of time. In Figure 121 we have redrawn a spherical universe that expands from a point to some large radius, then contracts back to a point. The important idea here is that there is no motion in this picture. That is, we are not intended to imagine a circle that starts out around the South Pole and slides up, becoming the Equator, the North Polar ice cap and finally a point again. Instead we are to think of this sphere of space-time as simply *existing*. In Hermann Weyl's words, "The objective world simply *is*, it does not *happen*." Asking what happens before the beginning of the universe here is a little like asking what continents are south of Antarctica. There is no time and space except the one we inhabit.

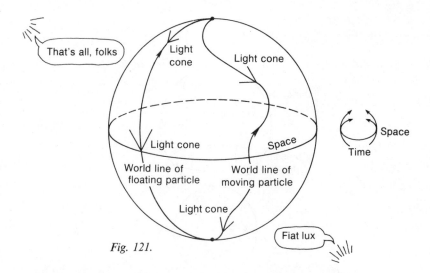

Fig. 121.

Nevertheless, this picture is rather unsatisfying for reasons of conservation of matter. What happens to all those particles that end up at the North Pole? Where did all those particles at the South Pole come from? One solution would be for there to be equal amounts of matter and antimatter in the universe. What would happen at the

North Pole would be mutual annihilation of particles; what would happen at the South Pole would be pair production. Taking the Feyman view of antimatter, we would have world lines that went from South to North as electrons and came back on the other side as positrons, forming a closed curve like a longitude line.

A different tack is to claim that there is always another universe after this one, and that there was always one before this one. This is the oscillating-universe or "string-of-pearls" model.

Each cycle of the universe is represented as a sphere. Space is curved, a 1-D circle in Figure 122, the 3-D hypersurface of a hypersphere in reality. Time is curved so that space expands and contracts. Let us emphasize that all these drawings have been for Lineland, 1-D space. The "pearls" on this string are like the sphere in Figure 121; each of them should really be a hyperhypersphere whose hyperhypersurface could serve as our 4-D space-time.

Notice that each cycle of the universe is different. It is speculated that every physical constant could come out different each time

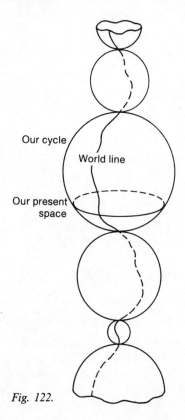

Fig. 122.

that all the universe's matter was squeezed through the "knothole" between two cycles.

This model makes the singularity of the beginning and end points of each cycle less striking, but only at the cost of reintroducing a cosmic time that goes on forever in both directions. We will now present a model (Figure 123) that avoids this unpleasantness and provides a satisfying answer to the questions of what came before and what comes after.

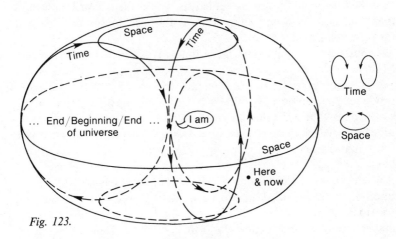

Fig. 123.

"Torus" is the mathematical word for donut, so we call this model toroidal space-time. This model can be obtained by taking the spherical space-time in Figure 121 and pushing down on the North Pole and up on the South Pole until they become the same point. This 2-D surface has one space and one time dimension. If our universe had toroidal space-time, we would actually require the surface of a hyperhypertorus.

In this model we have hyperspherical space that expands from a Big Bang, which it later contracts back to. Since our space is still expanding, we would be located perhaps at the point labeled Here & Now. You might be tempted to ask the question: "I can see that space is like a circle that keeps cycling down out of the hole, up around the donut and back into and through the hole. I wonder how many times it has already done this?" This is exactly the wrong question to ask! There is no last time around or next time around, because *nothing is moving*. Space-time is a 4-D manifold with a certain structure. It is there timelessly. We *feel* that we are going through time, but this is an *illusion*.

What would happen if you tried to get into the past in the

toroidal universe by going forward in time, through the hole and up around to stop at some time before your departure? There's no problem with such a trip taking too long, since, as we discussed in the last chapter, if you traveled close enough to the speed of light the trip could take as little of your time as you like. The problem here would be that when you went through the point labeled " . . . End/Beginning/End . . . of universe" you would die. Why? Because at this point space is contracted to a point, which means that you would get squashed, as would your molecules, as would their atoms, as would their particles, as would everything. Only pure energy can make it through this singular point.

But what would cause space-time to curve in the first place? Matter.

According to Einstein's General Theory of Relativity, matter produces space-time curvature. Freely falling particles travel along world lines that are time-like geodesics of space-time. Since space-time is curved by matter, its geodesics are curved, and hence one finds that particles near massive objects travel along curved world lines.

The cause of the curving of world lines near massive objects has traditionally been called "the force of gravity." According to General Relativity there is no "force" as such, simply a curving of space-time that makes it natural for the world lines to curve. General Relativity thus provides us with an essentially geometric explanation of gravitation.

What sort of space-time curvature, exactly, does matter produce? Let us imagine a massive line segment in Lineland. Forget about time for a minute and say we are looking down on Lineland from the direction usually called Future in our space-time diagrams. Then the space of Lineland might look like Figure 124.

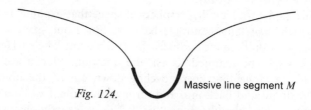

Fig. 124. Massive line segment *M*

If we had a number of such massive segments distributed in the space of Lineland, they could bend this space into a closed curve, approximately circular (Figure 125). Whether or not *our* space is actually closed—that is, curved into an approximate hypersphere—depends on how much matter our universe contains!

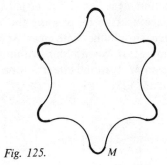

Fig. 125.   M

Now, what about the bending of the space-*time* of Lineland owing to the presence of matter? We have two experimentally observed facts to go on: (i) if a ruler is taken near a gravitating mass it appears shorter, and (ii) if a clock is taken near a gravitating mass it seems to run slower.

Say now that we consider the world line of the midpoint of a massive line segment M in Lineland. Fact (i) from above tells us that we should *stretch* the space coordinates near world line M, so that a ruler moved near M will look shorter (Figure 126).

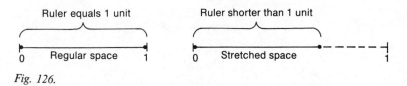

Fig. 126.

Fact (ii) from above tells us that we should *shrink* the time coordinates near world line M, so that the ticks of a clock moved near M will look longer (thus causing the clock to seem to tick slowly; Figure 127).

One way to accomplish these two objectives of stretching the

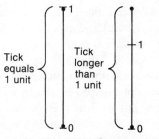

Fig. 127.   Regular time   Shrunken time

space and shrinking the time near world line M is as follows.

First stretch the space near M by pushing the world line of M away from us. And then shrink the time near M by bending the whole plane away from us. If we do this without stretching the vertical lines far away from M, then the effect will be to compress the vertical lines near M (Figure 128).

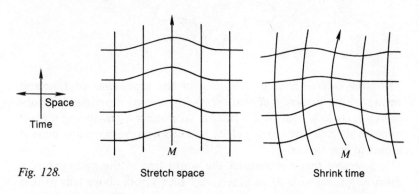

Fig. 128.   Stretch space   Shrink time

So one might expect the space-time of a Lineland with plenty of matter to be that of an expanding circular universe (Figure 129).

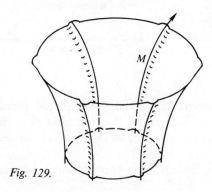

Fig. 129.

Indeed, Einstein suspected, on the basis of his calculations, that our space is both hyperspherical and expanding—and this was *before* the expansion of our universe had been detected by the astronomers.

Notice that the piece of Lineland's space-time that we have drawn in Figure 129 fits nicely into the picture of toroidal space-time given earlier. One could take Figure 129 and fit it into the hole of the space-time donut without any difficulty. By looking at the space-time donut one can also see how the expansion of the universe caused by the shrinking of the time axis near massive objects could eventually turn into a *contraction* of space.

Do these pictures *prove* that our space-time is a grooved donut, as I suggest? I wish they did, but pictures can be misleading. A real scientific proof must proceed analytically from assumptions that have been made explicit, both so that the plausibility of the assumptions and the correctness of the reasoning can be examined, and also so that testable quantitative predictions can be extracted from the theory.

General Relativity is analytically formulated in terms of a G-tensor like that of Chapter 3. In flat space-time with a standard coordinate system, the interval dI between the points with coordinates (x, y, z, t) and $(x + dx, y + dy, z + dz, t + dt)$ is given by the equation $dI^2 = dt^2 - dx^2 - dy^2 - dz^2$. In a curved space-time with a reasonable coordinate system it is possible to have this equation hold at some points, but not at all points, for if it held at all points the space-time would be flat.

In general we will only have

$$
\begin{aligned}
dI^2 = g_{11}\, \delta x^2 &+ 2g_{12}\, dx\, dy + 2g_{13}\, dx\, dz + 2g_{14}\, dx\, dt \\
&+ g_{22}\, dy^2 \quad\;\; + 2g_{23}\, dy\, dz + 2g_{24}\, dy\, dt \\
&\qquad\qquad\quad\;\; + g_{33}\, dz^2 \quad\;\; + 2g_{34}\, dz\, dt \\
&\qquad\qquad\qquad\qquad\qquad\quad\; + g_{44}\, dt^2.
\end{aligned}
$$

The value of each g_{ij} depends on the particular (x, y, z, t) you are working near. One usually thinks of the ten g_{ij} functions expressed by a single tensor-valued function $G\,(x, y, z, t)$. In flat space-time, of course,

$$
G(x, y, z, t) = \begin{bmatrix} -1 & 0 & 0 & 0 \\ 0 & -1 & 0 & 0 \\ 0 & 0 & -1 & 0 \\ 0 & 0 & 0 & 1 \end{bmatrix} \quad \text{everywhere.}
$$

Historically, the hardest part of formulating the General Theory of Relativity was to find the "field equations" specifying the G-tensor in terms of the distribution of mass and energy in space-time. Once you have the G-tensor, then you can determine the interval corresponding to any path in space-time by integrating (adding together) the dI's along the path in question; and you can decide which paths are geodesic, i.e., *straightest*.

There are three kinds of geodesics in space-time: space-like, light-like and time-like. A space-like geodesic is determined, as you would expect, by the condition that the space-like interval along it be minimal. A light-like (null) geodesic is determined by the natural condition that the interval along it is zero. It is perhaps surprising, however, that in space-time the interval along a *time*-like geodesic must be *maximal*. These three conditions can be shown to be consequences of the definition of "geodesic" as "straightest path."

In Lineland it turns out that a time-like geodesic of a particle P moving near the massive segment M would look like Figure 130 (if the particle were free to move through M). This world line is, of course, that of a particle oscillating back and forth. By staying near M, the world line gets "more time" (since the time scale is shrunken near M) and "less space" (since the space scale is stretched near M), and thus interval ($= \sqrt{\text{time}^2 - \text{space}^2}$) is maximized.

We can't draw the curved space-time of Flatland with a massive object (since this would require four dimensions), but it turns out that the time-like geodesic of a freely falling particle E near a massive particle S would look like Figure 131 in the space-time of Flatland.

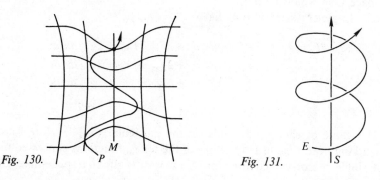

Fig. 130.

Fig. 131.

This picture can be thought of as representing the Earth's motion around the Sun. According to the general relativistic view of things, the Earth does not move around the Sun because of gravitational force, but rather because by moving around the Sun the Earth manages to maximize the interval along its world line. Again, the Earth "wants" to maximize the interval along its world line since it is moving freely through curved space-time and thus follows a world line that is as straight as possible. (Recall that the straightest time-like path can be mathematically proved to maximize interval.)

How can we see that the interval along the Earth's world line *is* maximal? The idea is that since the Earth is in free fall it can (by a generalization of the Principle of Relativity that is called the Equivalence Principle) regard itself as motionless. Now if Earth is motionless, then anyone who flies away from it and rejoins it later will be perceived by Earth to be in motion. But recall that if someone is in motion relative to Earth, then his clocks run slower than Earth clocks. So the time interval the traveler measures between leaving and coming back would be less than the time interval the Earth measured. Therefore Earth's time-like interval is maximal, so its path must indeed be a time-like geodesic in curved space-time.

It turns out that the world lines of light rays—null geodesics —are also curved in the presence of matter. This "light has weight" prediction of the General Theory of Relativity was first tested during the solar eclipse of 1919 (see Eddington's book in Bibliography). What is more exciting, however, is the fact that the path of a light ray that emanates from a dense enough star will be so curved that it falls back into the star, thus rendering the star invisible to us (Figure 132).

Fig. 132.

Such invisible stars are called black holes. Nothing can escape from a black hole. How exactly do they arise? Let us return to Lineland. The space near a dense segment *M* looks like Figure 133, as we said before.

Normally a star does not collapse under the force of its own gravitational attraction because this attraction is counterbalanced by the tendency to expand that a hot volume of gas has. But if a star cools it can contract and become denser (Figure 134).

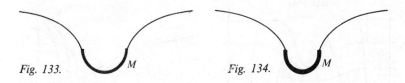

Fig. 133.   M   *Fig. 134.*   M

Normally the electric repulsion that the electrons of a star have for each other will keep it from collapsing further, but if it is massive enough this repulsion is overcome and the contraction continues (Figure 135).

Once a star has contracted beyond a certain point there seems to be nothing that can prevent it from actually contracting to a point. The stretching of space and shrinking of time near such a point tend to infinity, and the laws of physics break down at such a point, which is therefore called a *singularity* (Figure 136).

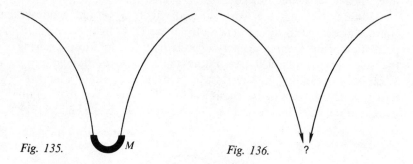

Fig. 135.   M   *Fig. 136.*   ?

We cannot hope to observe the singularity that a black hole evolves into, because once the contraction of the star has gone far enough its appearance from the outside does not change. No more signals of any kind escape from the star once it gets smaller than what is called the *event horizon*. Note, however, that if it were not for the fact that the strong field would crush one out of existence one could in principle travel to the singularity in a finite interval of time. For although the stretching of space becomes infinite, so that the singularity appears infinitely far away, as one approaches it the shrinking of time becomes infinite so that one's life becomes infinitely long. These infinities are only relative to the space-time outside the event horizon, however. The rash explorer of the black hole would find that he reached the singularity after what seemed to be only a few hours of *his* time.

Figure 137 shows two drawings of the space-time of a Lineland

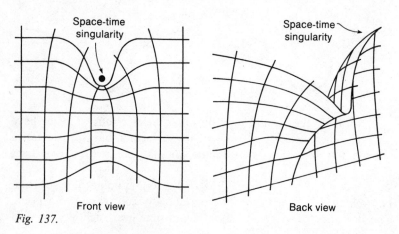

Space-time singularity   Space-time singularity

Front view   Back view

Fig. 137.

containing a dense segment that collapses to singularity. The line of
simultaneity for any individual in Lineland is bent to go under the
singularity, for an external observer views the full collapse as taking
forever. This is clear from the front view. From the back view one
can see that the distance to the singularity becomes infinite as well.

If one looks at these pictures and recalls that a light ray always
takes a world line that bisects the angle between the time axis and the
space axis, then it is also evident that light rays emitted near the
singularity cannot escape the "trough."

What actually happens at the singularity? If one has a hyperspherical space, then the view that all the singularities are the same
point is quite attractive. We have drawn in Figure 138 a Lineland
with circular space, and with two stars that have collapsed to singularity.

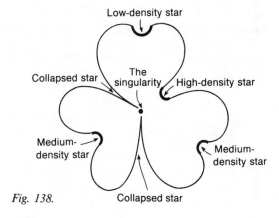

Fig. 138.

If it were only possible to survive going through a singularity one
could travel from one side of the universe to the other quite rapidly
by flying into a black hole and expecting to come out of a different
one.

It has actually been suggested that if a star is rotating rapidly
enough when it collapses, then it may be possible to fly into such a
black hole and emerge unscathed. Suppose there were a rotating
black hole into which stuff was falling, and emerging someplace else.
What would one call the place where all the stuff came welling up? A
white hole, naturally. There is speculation that each galaxy has a
white hole at its center, so that a galaxy is something like the
spreading puddle around a mountain spring.

If we go back to the space-time donut, this whole scenario can
be envisaged in space-time. The black holes will be grooves in the
donut that become deeper and go around over the donut to the

central singularity. White holes will be deep grooves that come directly out from the central singularity to join the surface and flatten out as they move on.

So it could be that all of the known space-time singularities—initial, final, black hole and white hole—are the same. What should this one singularity be called? No name seems quite adequate.

We started out this chapter by considering the structure of the universe in terms of the large-scale structure of space-time. We then showed how General Relativity explains gravitation in terms of medium-scale curvatures of space-time. It has been suggested by certain authors that the existence of *matter* is to be explained in terms of the small-scale curvature of space-time. In 1870, the mathematician William Clifford put it this way (see Misner, Thorne and Wheeler for reference):

> I hold in fact (1) that small portions of space *are* in fact of a nature analogous to little hills on a surface which is on the average flat; namely that the ordinary laws of geometry are not valid in them; (2) that this property of being curved or distorted is continually being passed on from one portion of space to another after the manner of a wave; (3) that this variation of the curvature of space is what really happens in that phenomenon which we call the motion of *matter*, whether ponderable or ethereal; (4) that in the physical world nothing else takes place but this variation, subject (possibly) to the law of continuity.

A slightly different view is that mass particles are actually tiny black holes, event horizons around a singularity. If, again, all singularities were the same point, then all of space-time, and all matter as well, would simply be a maze of "grooves" connected to the central singularity.

Again, it has been suggested that our space actually has a slight 4-D thickness to it, and that the elementary particles are small hyperspheres constrained to move in our hyperplane. This slight four-dimensionality of our space might be noticeable in the phenomenon of two particles moving toward each other on a direct collision course, but somehow missing each other (compare Figure 139).

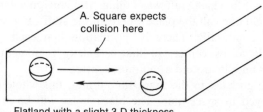

Fig. 139.   Flatland with a slight 3-D thickness

What would these little hyperspheres be made of—pure curved space? Perhaps, but another interesting possibility is that each of these little hyperspheres is somehow identical with the large hypersphere which is our space. This notion is developed in my novel, *Spacetime Donuts*.

The best thinking these days on the question of the small-scale structure of space-time seems to be that there is really no unique structure once you go down past the 10^{-33} centimeter size level (see the last chapter of Misner, Thorne and Wheeler). The idea is that space-time is "foam-like" at this size scale with connections between widely separated events continually forming and disappearing. Wheeler has even suggested that the phenomenon of electric charge can be explained in terms of such a multiply connected space-structure.

PROBLEMS ON CHAPTER 7

(1) Consider these two different kinds of expanding universes: *Flat Space:* 12 billion years ago all the matter in the universe was concentrated at one point in space. There was a huge explosion, and the fragments of this explosion have been hurtling through space away from the explosion point ever since. *Spherical Space:* 12 billion years ago the radius of our hyperspherical space was zero. Space was a single point with infinite energy density. There was a huge "explosion" and space began expanding, carrying fragments of matter with it.

Now answer these questions for the two models: (a) Where is the space location of the beginning of the universe? (b) Is it possible for us to see (i.e., to receive light signals from) the beginning of the universe? (c) If the universe started contracting would there be any way to avoid collapsing back to a point along with the rest of the matter in the universe?

(2) If, as in the drawings we made of Lineland's space-time, we visualize a bit of matter as being a groove in space-time, how should one visualize a bit of antimatter, given that when matter and antimatter meet they both disappear?

(3) A toroidal space-time can be obtained by taking a tubular section of the circular-time universe described in Chapter 6 and joining the

two ends together with the same orientation (Figure 140). In 6-D space it would be possible to join the two ends together with the time circles having *opposite* orientation, to produce a "Klein bottle space-time" (see the end of Chapter 3). What would you find upon returning from a journey around the circular space of such a universe? Would it be possible to fit the individuals' times into a consistent universal time in such a universe?

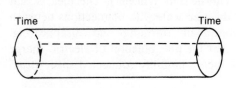

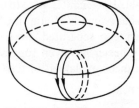

Fig. 140.

Two time directions match
in toroidal space-time

(4) The Twin Paradox of Special Relativity goes like this: "Say that my twin brother flies away from Earth at almost the speed of light for ten years, then stops his rocket and flies back in ten more of his years. When he returns I will have aged 20 years and he will have aged, say, one year. Could he not argue that *I* moved away and back to him, so that *my* biological clock ran slower, and thus expect *me* to have aged 19 years less than him?"(Figure141). The answer is no, since when the traveler turns around he shifts reference frames (see Taylor and Wheeler for details). But what if space were hyperspherical so that the traveler would never have to " turn around "?

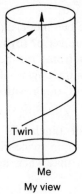

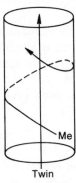

Fig. 141.   My view   Twin's view

8

CONCLUSION

Geometrical objects are unchanging forms. The goal of this book has been to present the universe as a geometrical object that happens to enjoy the property of being perceived by us to exist.

We showed how the passage of time and the apparent changes in our universe could be eliminated by thinking in terms of 4-D space-time.

The relativity of simultaneity, more than anything else, forces the view that time is not really passing. This argument is presented in Gödel's paper, "A Remark about the Relationship between Relativity Theory and Idealistic Philosophy," in the Schilpp anthology. The idea is that if simultaneity is a relative concept, then it is impossible to think of space-time as being a stack of unique "nows" that successively appear and vanish from existence. The past and future really exist.

This is a valuable thing to keep in mind. As John Updike writes in *Rabbit Redux,* "Time is our element, not a mistaken invader." It is a mistake to look forward to the good times and fear the bad times. They are all part of the 4-D object which is you. The best way to get close to the Eternal is to get close to the Now, for there is no time right now. Time arises when we grasp at the world with our rational mind.

Once we accept the 4-D viewpoint it is possible to see the universe as a single object whose structure we can investigate. There is some hope that every feature of the universe can be reduced to some geometrical property of the space-time manifold. Einstein made the decisive first step in this task when he reduced gravitation to curvature of the space-time manifold.

To see the universe as a single object is a great thing. We will conclude with a quote on this from one of Einstein's letters:

A human being is a part of the whole, called by us "Universe," a part limited in time and space. He experiences himself, his thoughts and feelings, as something separated from the rest—a kind of optical delusion of his consciousness. This delusion is a kind of prison for us, restricting us to our personal desires and to affection for a few persons nearest to us. Our task must be to free ourselves from this prison by widening our circle of compassion to embrace all living creatures and the whole nature in its beauty. Nobody is able to achieve this completely, but the striving for such achievement is in itself a part of the liberation and a foundation for inner security.

ANNOTATED BIBLIOGRAPHY

Edwin A. Abbott, *Flatland* (Dover reprint, New York, 1952).

The one and only. *Flatland* first appeared around 1884. Like so many of the writers on these topics, Abbott was an Englishman, sharing in that national resource of solemn jocosity. *Flatland* is a very funny book.

It is written in the form of A. Square's autobiography, and has two parts. Part II contains A. Square's dimensional adventures involving Pointland, Lineland, Flatland, Spaceland and Thoughtland. Part I contains very little of this and is more Swiftean social satire than anything else.

For instance, despite the fact that he is writing his memoirs in prison, A. Square can still recount the Flatland government's ruthless treatment of the Irregular Polygons (the Flatland equivalent of cripples) with Toryish approval: "Let the advocates of a falsely called Philanthropy plead as they may for the abrogation of the Irregular Penal Laws, I for my part have never known an Irregular who was not also what Nature evidently intended him to be—a hypocrite, a misanthropist, and, up to the limits of his power, a perpetrator of all manner of mischief. ... I would suggest that the Irregular offspring be painlessly and mercifully consumed."

In Flatland, the more sides a regular polygon has, the greater is his social standing. Women are line segments. A. Square's remarks on women are provoking enough to have aroused comment even in the nineteenth Century; so much so that in the Preface to the Second Edition (purportedly written by one of A. Square's acquaintances) it is reported that: "It has been objected that he [A. Square] is a woman-hater; and as this objection has been vehemently urged by those whom Nature's decree has constituted the somewhat larger half of the Space-land race, I should like to remove it, so far as I can honestly do so. But the Square is so unaccustomed to the use of the moral terminology of Spaceland that I should be doing him an injustice if I were literally to transcribe his defence against this charge."

One aspect of A. Square's Spaceland experiences in Part II seems to be especially problematic. This is his ability to see 2-D images from Spaceland. Thus, for example, he can look down on his house and see the inside of every room, and the inside of the bodies of his sleeping family. Certainly a 3-D person with a 2-D retina would see this, but A. Square's retina is presumably 1-D, a line segment at the back of his flat eye.

The point I am making is that it would seem that even if lifted above Flatland, A. Square's visual space would continue to be a plane (that of his body) collapsed into a line (the retinal image) with variations of brightness. In the same sense, if we were lifted into hyperspace, our visual space would continue to be a 3-D space (that of our body) collapsed into a plane (the retinal image) with variations of brightness. So if A. Square, floating above Flatland, were to rotate back and forth he could see every possible 1-D cross section of Flatland, and perhaps put these together to form the full 2-D image, but he would not *see* the whole 2-D Flatland at once. In the same way, if we were to enter hyperspace we could look at our space, and as we moved about see every possible 2-D cross section of it, perhaps mentally combining these images to form the full image (inside and out) of everything in our space. Of course, if, upon being whisked into hyperspace, we were equipped with an astral body complete with 4-D eyes, this problem would not arise.

Jorge Luis Borges, *A Personal Anthology* (Grove Press, New York, 1967). For our purposes, the interesting essay in here is "A New Refutation of Time." However, the other stories and essays herein are also quite fascinating. "The Aleph," for instance, could be viewed as a description of how a space-time singularity might look. Borges has written a number of other scientifically interesting stories. "The Garden of Forking Paths," from *Ficciones*, for instance, furnishes the epigraph of the DeWitt book on branching time.

In "A New Refutation of Time," Borges takes the metaphysical idealism of Berkeley and Hume to what seems to be its necessary logical conclusion: "I deny the existence of one single time in which all events are linked. . . . Each moment we live exists, not the imaginary combination of these moments." So there are only separate mental states, conventionally tied together into a time stream. To make clear the way in which the time stream is disrupted by a full idealism, Borges reasons thus: "We can postulate, in the mind of an individual . . . two identical moments. . . . Are not these identical moments the same moment?" (It is interesting to note here that Kurt Gödel, who has also written on the unreality of change, uses a similar argument to support the view that thoughts have a reality external to us . . . since two different people can have the same thought.)

This idealistic destruction of time is illustrated by a beautifully written example. Borges later closes with a terribly sad paragraph beginning, "*And yet, and yet* . . . To deny temporal succession, to deny the self, to deny the astronomical universe, are measures of apparent despair and of secret consolation," and ending, "The world, unfortunately, is real; I, unfortunately, am Borges."

Claude Bragdon, *A Primer of Higher Space* (Omen Press, Tucson, Arizona, 1972).

This lovely little book was originally published in Rochester in 1913. Bragdon was a remarkable individual who not only worked as an architect and scenic designer but also wrote some 17 books. He was tied in with many of the mystical and occult movements of his time, and his book on mysticism and the fourth dimension, *Explorations into the Fourth Dimension* (originally, *Four Dimensional Vistas*), was reprinted in 1972 by the CSA Press in Lakemont, Georgia.

Dionys Burger, *Sphereland* (Thomas Y. Crowell Co., New York, 1965; Apollo Editions, New York).

Originally written in Dutch, *Sphereland* starts with a summary of Abbott's *Flatland* and then goes on to an account purportedly written by A. Square's grandson, A. Hexagon.

This book lacks the satiric bite of *Flatland*, but it provides a nice dramatic account of how the Flatlanders might discover the curvature of their space (into a sphere) from the fact that the sum of the angles in a sufficiently large triangle is appreciably larger than $180°$; and of how the Flatlanders might come to explain the observed recession of distant objects by viewing their world as the surface of an expanding sphere. An odd feature of this book is that here the Flatlanders are somewhat like birds, that is, they live in a disk (compare: spherical atmosphere) in the center of which is an attracting mass. Their natural paths are circles around this central mass.

Carlos Castaneda, *A Separate Reality* (Simon and Schuster, New York, 1971).

Carlos Castaneda has written a series of four books about his encounters with a Mexican *brujo* or sorcerer called Don Juan. *A Separate Reality* is the second in the series, and perhaps the best.

The basic idea behind Don Juan's teachings is that we create the world around us by our assumptions. Our rational system of interpretation carves out a certain set of perceptions, connects them in a certain way and announces, "The world is like this." Don Juan bears this in on Carlos by forcing him to let down his guard and interpret reality in radically different ways. Specific ways of achieving this are described: the ritual use of psychedelics, stopping one's internal monologue, concentrating on sounds rather than sights and attempting to wake up inside your dreams.

Don Juan's goal is not so much to win Carlos over to a belief in spirits, talking coyotes, etc., as to show him that such beliefs are as reasonable as belief, say, in the content of the 6:30 news. What you see depends on what you are prepared to see.

I used to think that Don Juan was trying to get Carlos to start seeing things in terms of space-time. One incident, in particular (where Don Juan gets Carlos to see the same leaf fall off a tree three times in a row), seems to support this view. But more recently I have come to think that Don Juan's real teaching is that it is possible, when we realize that all systems of interpretation are equally arbitrary, to leave this one possible world and live for a time in what Wheeler calls Superspace.

One would think that viewing all possible worlds as equally valid would destroy any justification for ethical considerations, but Don Juan answers this problem: "I choose to live, and to laugh, not because it matters, but because that choice is the bent of my nature. . . . A man of knowledge chooses a path with heart and follows it. . . . Nothing being more important than anything else, a man of knowledge chooses any act, and acts it out as if it matters to him" (pp. 106–107).

Bryce S. DeWitt and Neil Graham, editors, *The Many-Worlds Interpretation of Quantum Mechanics* (Princeton University Press, Princeton, N.J., 1973).

The heart of this paperback is Hugh Everett's monograph, "The Theory of the Universal Wave Function." There is also, among other things, an "Assessment" by John Wheeler, as well as an elementary presentation of Everett's theory by Bryce DeWitt.

Everett's starting point is that in quantum mechanics a system can change in two ways (given here in reverse order). *Process 2:* When a system is left on its own it is not definitely in any one eigenstate. The probabilities of its being observed in various eigenstates evolve in a continuous way as time goes on. *Process 1:* When a measurement is performed on a system a discontinuous "collapse of the state vector" occurs, so that the probability of one particular eigenstate becomes 1 and that of all the others becomes 0.

Now say that you (or a cat) are in a room and you make a measurement on a certain system at noon. You feel that the state vector of the system collapses then. But for someone outside the room who views you-plus-the-system-you-are-observing as a single larger system, your results do not collapse into a unique eigenstate until he opens the door and looks at you.

So, Everett suggests, since someone else can always walk in, why not assume that *we* are also state vectors, existing in many different eigenstates at once with certain probabilities, and that "one of us" observes every possible outcome of every experiment we conduct. The image here is that of a branching universe, although actually the branching is so thick that a continuous Superspace of possible universes is a more appropriate image.

If the universe is really like this, is there any way for you to ensure that the next time you take an airplane you do not go into the "branch" in which the airplane crashes? Somehow it is not enough to simply say that you split and that a "you" goes into both of the possible futures, for one has the compelling feeling that the life one *actually* lives is different from the other possible lives. Of course, this feeling can be written off as an illusion on a par with one's illusion that time really passes, but there is a nagging feeling that there is something more to it.

The question just raised is equivalent to the question of how this model can correctly explain the observed probability distributions.

J. W. Dunne, *An Experiment with Time* (Faber and Faber, London, 1969).
This curious book was originally published in 1927. On the basis of certain memorable experiences of what seemed to be dream precognition, Dunne had come to believe that our dreams draw upon future as well as past events. In this book he describes his attempts to prove that dream precognition actually occurs, as well as his theory (Serial Time) of how such a phenomenon could be possible.

The idea behind the experimental tests was to get a large group of people to write down all the specific dream images that they could recall each morning, and then to keep an eye out for the realization of these images. A number of Dunne's subjects did observe such dream precognition in themselves, but, as Dunne himself points out, the greater number of these observations can be written off to chance or autosuggestion. Nonetheless, the idea is an interesting one and it has been, for instance, my experience that if you begin looking for this phenomenon you will find rather striking instances of it.

Dunne's Serial Time theory carries what David Park has called "the fallacy of the animated Minkowski diagram" to its logical conclusion. That is, Dunne starts out with his world line in space-time, but then asserts that his consciousness is *moving* along this world line. Of course the *time* in which his consciousness is *moving* is different from the frozen physical time of the space-time block universe. (Similarly, in Vonnegut's *Slaughterhouse Five*, the hero's consciousness moves in a time outside that of physical space-time.) Now, since the consciousness exists outside of space-time, it can range freely over it, picking a bit here and a bit there to weave its dreams. Moreover, if the consciousness discovers something unpleasant in the future of physical space-time as it stands, it can alter this space-time. We thus have a space-time that is changing while a second time (the time of consciousness) lapses.

This whole train of thought can be repeated for the frozen space-time of the world and the consciousness, so that one ultimately has an infinite regress of times and consciousnesses. As I have indicated in several places, I subscribe to the view that our sensation of the *passage* of time is to be viewed as an illusion, an artifact of the space-time geometry of the universe. Nevertheless, Dunne's working out of his Serial Time notion is quite interesting, and could perhaps be viewed as a vague precursor of the notion of Superspace described in Misner, Thorne and Wheeler.

One phrase sticks in the mind, "Consequently, you, the ultimate, observing you, are always outside any world of which you can make a coherent mental picture."

Arthur S. Eddington, *Space, Time and Gravitation* (Harper & Row, New York, 1959).

This book was first published in 1920. Eddington wrote a number of books of what one might call either popular science or philosophy of science, and they are all excellent. One does not soon forget Eddington's pleasant style, relaxed, slightly humorous, but totally serious; *engagé*, but never partisan.

Space, Time and Gravitation begins with a prologue on the nature of geometry that brings out vividly the problem of what exactly it means to say that the length of a ruler does or does not depend on where in the universe it is located. The book then moves through a clear description of special relativity in terms of Minkowski diagrams (perhaps the first such popular presentation), a fairly detailed exposition of general relativity and a description of the experimental tests of general relativity. Chapter 11 is unique in that it gives the only available popular presentation of H. Weyl's geometric theory of electromagnetism in terms of the "gauge." (See, however, Weyl's rather difficult book, *Space, Time, Matter.*) *The Mathematical Theory of Relativity*, Eddington's companion piece to *Space, Time and Gravitation*, was reprinted in 1975 by Chelsea. To a certain extent, the latter was written as an introduction to the former, and the interested reader may wish to use it this way.

The last chapter of *Space, Time and Gravitation* presents Eddington's striking idea that "where science has progressed the farthest, the mind has but regained from nature that which the mind has put into nature." That is, "Our whole theory has really been a discussion of the most general way in which [the illusion of] permanent substance can be built up out of relations; and it is the mind which, by insisting on regarding only the things that are permanent, has actually imposed these laws on an indifferent world." He develops this idea further in *The Philosophy of Physical Science* (popular) and took it to an extreme in *Fundamental Theory*. In this last work (which was posthumously assembled from his notes), Eddington sets out to derive all of nature's physical constants (e.g., Planck's constant, the mass of the electron, the radius of the universe, etc.) on the basis of certain *a priori* epistemological considerations. This book represents one of the earlier attempts to wed general relativity to quantum mechanics, "a fiery marriage which has yet to be consummated," in the words of John Wheeler.

Albert Einstein, *Relativity: The Special and the General Theory* (Crown Publishers, New York, 1961).

The body of this little book is a translation of a popular exposition written by Einstein in 1916. There is also a most interesting Appendix, "Relativity and the Problem of Space," written in 1952.

Most popular expositions of relativity theory are in fact rehashes of this beautiful book. It is, however, virtually impossible to improve upon Einstein's clear and friendly presentation. The Appendix mentioned above makes the points that "It appears therefore more natural to think of physical reality as a four-dimensional existence, instead of, as hitherto, the *evolution* of a three-dimensional existence," and that "Space-time does not claim existence on its own, but only as a structural quality of the field," and ends with a brief description of Einstein's attempts to arrive at a unified field theory.

Albert Einstein, *Sidelights on Relativity* (E. P. Dutton and Co., New York, 1923).
Unfortunately, this highly readable book is out of print and not easy to get hold of. It contains translations of two of Einstein's addresses, "Ether and the Theory of Relativity" (1920) and "Geometry and Experience" (1921).

Reading some popularizations, one gets the impression that relativity did away with the "ether" which classical physicists had supposed to fill the spaces between particles and provide the medium for the transmission of light. In his first address, Einstein makes clear the limited sense in which this is correct: "The special theory of relativity forbids us to assume the ether to consist of particles observable through time, but the hypothesis of ether in itself is not in conflict with the special theory of relativity." He then goes on to describe the way in which the general theory of relativity is, in effect, an ether theory: "The recognition of the fact that 'empty space' in its physical relation is neither homogeneous nor isotropic, compelling us to describe its state by ten functions (the gravitation potentials $g_{\mu\nu}$), has, I think, finally disposed of the view that space is physically empty. But therewith the conception of the ether has again acquired an intelligible content."

The address, "Geometry and Experience," is highly relevant to my book. Consider this quote: "Geometry must be stripped of its merely logical-formal character by the co-ordination of real objects of experience with the empty conceptual framework of axiomatic geometry. . . . Geometry thus completed is evidently a natural science; we may in fact regard it as the most ancient branch of physics." Having made this remark, Einstein goes on to consider what might be the geometry of the universe. In order to avoid Poincaré's conventionalistic view that the world is Euclidean and that any non-Euclidean behavior of physical objects can be ascribed to various "forces," Einstein makes the explicit assumption that "If two tracts are found to be equal once and anywhere, they are equal always and everywhere." That is, rather than saying a meter-stick shrinks near a dense object, we say that the space near the dense object is stretched.

He ends this address with a description of the correspondence between the sphere and the "flat sphere" via stereographic projection, and the way in which one can thus visualize a hypersphere by imagining our space to be a "flat hypersphere."

Albert Einstein, *The Meaning of Relativity* (Princeton University Press, Princeton, 1953).
The body of this book, first published in 1922, consists of the text of

four lectures Einstein delivered at Princeton. Einstein provides us here with a compact and sophisticated development of special and general relativity.

An interesting answer to the question of what is reality can be found on the first page of this book: "By the aid of speech different individuals can, to a certain extent, compare their experiences. In this way it is shown that certain sense perceptions of different individuals correspond to each other, while for other sense perceptions no such correspondence can be established. We are accustomed to regard as real those sense perceptions which are common to different individuals, and which therefore are, in a measure, impersonal."

A. Einstein, H. A. Lorentz, H. Weyl and H. Minkowski, *The Principle of Relativity* (Dover Publications, New York, 1952).
This collection of translations of the original papers on the theory of relativity was first put out in 1923.

Einstein's first relativity paper, "On the Electrodynamics of Moving Bodies" (1905), is here, and the casual reader can expect to read the introduction and first two subsections without undue difficulty. It is a thrilling experience to make direct contact with the birth of relativity in this way.

Minkowski's famous paper, "Space and Time" (1908), is here as well, and I would urge the interested reader to go through at least the first two sections of this paper. It was Minkowski who invented the geometric interpretation of special relativity that I have used, and he presents it here with great clarity. Minkowski's style has a certain *panache* in this translation: "With this most valiant piece of chalk I might project upon the blackboard four world-axes. Since merely one chalky axis, as it is, consists of molecules all a-thrill, and moreover is taking part in the earth's travels in the universe, it already affords us ample scope for abstraction; the somewhat greater abstraction associated with the number four is for the mathematician no infliction."

J. T. Fraser, F. C. Haber and G. H. Muller, editors, *The Study of Time* (Springer-Verlag, Berlin, 1972).
This book contains the papers presented at the First Conference of the International Society for the Study of Time in 1969. The essay, "The Dimensions of the Sensible Present," by H. A. C. Dobbs, from which I took the idea of the Necker cube reversal being a 4-D phenomenon, appears here.

The most valuable essay in the book is "The Myth of the Passage of Time," by David Park. Park argues convincingly in support of the position that time does not really "pass"; that once you have drawn the Minkowski diagram of a space-time, nothing is gained by "animating" this diagram by imagining a particular spatial cross section that moves upward through the diagram while a second time elapses.

Martin Gardner, *Relativity for the Million* (Macmillan, New York, 1962).

This book is copiously and attractively illustrated. I first got hold of a copy when I was a junior in high school. Full of ideas and pleasant to read, it was this book that kindled my continuing hope to some day fully understand general relativity. This is certainly one of the best elementary treatments available.

Gardner's *The Ambidextrous Universe* contains some interesting material on 4-D space, and his collection entitled *Mathematical Carnival* has a section on hypercubes.

S. W. Hawking and G. F. R. Ellis, *The Large Scale Structure of Space-Time* (Cambridge University Press, Cambridge, 1973).

This beautifully illustrated technical work is a study of the properties of space-time (curvature, causality, etc.) and of what happens when you reach an "edge" of space-time, a singularity.

The authors discuss the fact that certain considerations imply "the existence of a singularity in the past, at the beginning of the present epoch of expansion of the universe. This singularity is in principle visible to us. It might be interpreted as the beginning of the universe." A number of interesting results about black holes are proved.

The average reader will not be able to follow the arguments in detail, but the illustrations are definitely worth looking at. This book is, for instance, the best place to go to learn about Penrose diagrams. On page 169 there is a particularly interesting sketch of Gödel's universe.

David Hilbert and Stephan Cohn-Vossen, *Geometry and the Imagination* (Chelsea, New York, 1952).

This is a translation of the 1932 elaboration by Cohn-Vossen of Hilbert's 1920 lectures in Göttingen.

Section 23 of this lovely book is the most relevant here. In this section the authors describe the six regular 4-D polytopes. Five of them are analogous to the five regular 3-D polyhedra, and one, the 24-Cell, has no analogue in other dimensions. In all *n*-dimensional spaces, for *n* greater than or equal to 5, there are only three regular polytopes, the analogues of the cube, tetrahedron and octahedron.

Geometry and the Imagination also has a fascinating and highly visual chapter on differential geometry, as well as a very thorough discussion of the Klein bottle and the projective plane as closed surfaces in 4-D space.

C. Howard Hinton, *The Fourth Dimension* (Sonnenschein, London, 1904).

In 1888, Swann Sonnenschein & Co. published a book by C. H. Hinton called *A New Era of Thought*. In this book Hinton suggests that our space may have a slight 4-D hyperthickness, so that the ultimate components of our nervous system are actually higher-dimensional, thus enabling human brains to imagine 4-D space. "The particular problem at which I have worked for more than ten years, has been completely solved," Hinton says. "It is possible for the mind to acquire a conception of higher space as adequate as that of our three-dimensional space, and to use it in the same manner." He then outlines in detail a series of mental exercises to be carried out with a set of 27 colored cubes which fit together into a single large cube (at one time Hinton's publisher actually sold sets of these cubes). The idea was that one can learn to know the intrinsic "next to" relations in a cube independent of any particular embedding of the cube in 3-D space; and if you can learn to think of a cube and its mirror image as the same thing, you are on your way to thinking in 4-D space.

Part of *The Fourth Dimension* is also devoted to these exercises and variants thereof. I cannot say that I have devoted much time to the exercises, as they seem unbearably tedious, involving the memorization of scores of arbitrary labels. There are, in my opinion, better and more direct ways of learning to "see" 4-D space.

However, the gradual path to enlightenment was manifestly efficacious in Hinton's case, as *The Fourth Dimension* contains a number of interesting insights into higher-dimensional space. For instance, there is a detailed analysis of the types of rotation which would be possible in 4-D space, leading up to a representation of electricity as a vortex ring in a 4-D ether. There is also Hinton's remarkable anticipation of the Minkowskian geometry which is based on *interval* instead of *distance*.

Perhaps the most interesting of Hinton's writings are those collected in the two volumes called *Scientific Romances*. The First Series, published by Swann Sonnenschein in 1904 contains, among other things, "A Plane World," which develops the physics which would obtain in Abbott's Flatland. The Second Series appeared in 1909, and here as always, Hinton insists upon the beneficial effects of the higher-space viewpoint: "And I have often thought, travelling by railway, when between the dark underground stations the lads and errand boys bend over the scraps of badly printed paper, reading fearful tales—I have often thought how much better it would be if they were doing that which I may call 'communing with space.'" This volume ends with a strange modernistic story called "An Unfinished Communication" which deals with a young man's experiences with an "Unlearner" and his subsequent detemporalization. This last story is somewhat reminiscent of P. D. Ouspensky's novel about a man who returns to his youth and finds himself compelled to repeat all of the mistakes he made, *The Strange Life of Ivan Osokin*.

William J. Kaufmann, *Relativity and Cosmology* (Harper and Row, New York, 1973).

This paperback popularization is an astronomer's discussion of recent theoretical and experimental developments in the field of cosmology. He cites some interesting experimental evidence for the propositions that our space is curved into the hypersurface of a hypersphere, and that space will eventually stop expanding and contract back to a singularity. This book also contains an intriguing discussion of the continuing puzzle of the quasars, as well as some relatively new material on black holes.

Henry P. Manning, editor, *The Fourth Dimension Simply Explained* (Dover Publications, New York, 1960).
Originally published in 1910, this book is a collection of some of the essays submitted to *Scientific American* when that magazine offered $500 to the author of the best popular explanation of the fourth dimension.

There is a great wealth of ingenious analogies and examples to be found here, generally along the lines of those I mentioned in Chapter 1. Manning provides a very comprehensive introduction, and is careful to point out the places where the essayists have made incorrect statements.

Charles W. Misner, Kip S. Thorne and John Archibald Wheeler, *Gravitation* (W. H. Freeman, San Francisco, 1973).
This book is heavy in every sense of the word. Some 1200 pages long, it describes Einstein's theory of gravity and the many modern tests and applications of this theory. If you want to get the inside dope on general relativity, black holes, cosmology or the like, this is the place to go.

Gravitation is quite advanced on the whole, but the authors do everything they can for the casual reader. There are scores of interesting boxes, figures and diagrams, and it is possible to skate through almost any section in the book with some comprehension.

The last two chapters are particularly interesting, as here the authors are working at the edge of their knowledge. The next-to-last chapter, "Superspace: Arena for the Dynamics of Geometry," presents the truly revolutionary idea that there is a continuum of possible universes at any time. A space-time or "leaf of history" seems to arise when a high-probability family of spaces fit together to make a space-time. But all the other, less likely, spaces exist as well, although it is not quite clear which possible space-times (other than the one we perceive) are "real."

The last chapter, "Beyond the End of Time," contains ideas mostly due to Wheeler. I thought of toroidal space-time as an answer to the question implicit in this chapter's title. There is a great wealth of far-out ideas in this chapter, and I urge you to read it.

Robert A. Monroe, *Journeys out of the Body* (Anchor Press/Doubleday, Garden City, N.Y., 1973).
So you're tired of just reading about 4-D space and want to go see it for yourself? This book tells you how to get there. Unfortunately, it is also a blueprint for insanity.

Monroe describes a fairly effective method of inducing a state in which one has the feeling of being able to leave one's body, move through walls and so on. Although he never refers to the fourth dimension, the idea of investigating the sort of "astral travel" he describes with an eye to interpreting the observed phenomena in terms of hyperspace is a tempting one.

The technique is basically to "wake up inside your dreams." It is not uncommon for one to have this experience during a daytime nap: that is, that one is awake and aware although one's body is still asleep. If one begins to look for this experience it begins to happen more often, and then astral travel is not far behind.

I worked on this for a few months once, but finally had to give it up as the experiences were so deeply frightening and disturbing. To be fully conscious and aware, and to know that one is in a dream world where anything can happen, to try to wake one's body up and not be able to—aaauugh! Indeed, reading the book, one gets the impression that Monroe finally scared himself into a heart attack.

But forewarned is forearmed, and perhaps some intrepid reader will be able to make something of the old theory that we have souls that move in hyperspace.

P. D. Ouspensky, *Tertium Organum* (Random House, New York, 1970).
This book was first published in 1922 and is now available as a Vintage paperback. Claude Bragdon was involved in its original English translation. Ouspensky wrote a number of other books. His *A New Model of the Universe* contains interesting chapters on "The Fourth Dimension" and on "Experimental Mysticism" (apparently about hashish). Ouspensky's book *In Search of the Miraculous* is, to my mind, the best available description of the teachings of G. I. Gurdjieff.

Tertium Organum is about the fourth dimension, and about a number of occult and mystical notions that can be thought of in terms of the fourth dimension. For instance, the distinct members of the human race can be thought of as being connected in a higher dimension, just as the separate cross sections of your fingers in Flatland are all part of your 3-D hand. It can be argued that the consciousness of a snail is 1-D, that of a horse 2-D, and that the goal of the mystic is to attain a consciousness that is 4-D.

Ouspensky's logic is occasionally faulty, but his basic notion of mystical consciousness as four-dimensional, both in the sense of timeless, and in the sense of seeing a higher unity above the world's diversity, rings true.

Robert L. Reeves, *Space and the Fourth Dimension* (Crescent Publishers, Grand Rapids, Michigan, 1922).
In which the author supplies his answer to the burning question: "Wherein is the Christian Scientist justified in making the assertion that this revelation of Truth which came to Mrs. [Mary Baker] Eddy is superior to Einstien's [sic] mathematical-physical deductions?"

Hans Reichenbach, *The Philosophy of Space and Time* (Dover Publications, New York, 1957).
This book was originally published in German in 1927. Reichenbach had a powerful imagination, and endeavored to "visualize" such things as non-Euclidean space, the fourth dimension, space-time and the hypersphere.

This book was most useful in the writing of *Geometry and Reality*. I am indebted to Reichenbach for some of the ideas in my Chapters 2 and 3, for the circular-time model mentioned in my Chapter 6 (this appears on page 272 of Reichenbach's book) and for the idea of using a torus to get a surface where a circle's expansion can smoothly turn into contraction.

A particularly interesting section of the book is called "The Number of Dimensions of Space." In this section, Reichenbach attempts to visualize a 4-D world by using color as the fourth dimension. That is, he asks us to think of a 3-D world in which objects pass through one another if their colors (i.e., 4-D locations) are different. In this section he also makes the speculation that the elementary particles might be tiny hyperspheres.

Wolfgang Rindler, *Essential Relativity* (Van Nostrand Reinhold Company, New York, 1969).

A college textbook, this work is one of the most accessible of the rigorous presentations of special relativity, general relativity and cosmology. The author has done original work on the paradoxes of special relativity (such as the pole-and-barn paradox) and his discussion of them is truly inspired.

The book's first chapter is fairly self-contained and gives us a very lucid examination of Mach's principle and its relationship to the equivalence principle of general relativity.

Paul Arthur Schilpp, *Albert Einstein: Philosopher-Scientist* (Harper and Row, New York, 1959; Open Court, Lasalle, Ill., 1973).

This consists mainly of essays on Einstein's work. It also includes a 45-page intellectual autobiography by Einstein (at the beginning) and his remarks on some of the included essays (at the end). These "critical remarks" are most interesting since they contain Einstein's reasons for refusing to accept quantum mechanics as a final physical theory.

One of the most important of the essays in this book is Kurt Gödel's "A Remark about the Relationship between Relativity Theory and Idealistic Philosophy." The aim of this essay is to show that the past and future exist statically and that time does not really pass. Gödel's first point is that, given the relativity of simultaneity, it is impossible to slice space-time into a stack of "nows" in any unique way, indicating that it is unrealistic to suppose that the world actually consists of such a series of fleeting "nows" with the past and future nonexistent. Gödel then goes on to describe an interesting model of the universe that he invented, in which "the local times of the special observers . . . cannot be fitted together into one world time." This happens because Gödel's universes (i) contain individuals whose world lines appear to us to be a pattern of simultaneous events, and (ii) admit the logical possibility of traveling back to points in one's own past. "Consequently, the inference drawn above as to the non-objectivity of change doubtless applies at least in these worlds." I have attempted my own example of such a world in Problem 3 of Chapter 7.

Hermann Schubert, *Mathematical Essays and Recreations* (Open Court Publishing Co., Chicago, 1903).
This is a translation of a book of essays that first appeared in the late 1800's. The essay that is relevant here is entitled "The Fourth Dimension."

"The Fourth Dimension" is primarily an attack on "Zöllner and his adherents," who had been claiming that spirits live in a 4-D space in which our space is embedded. This essay contains an interesting discussion of whether hyperspace really exists, and it is a rich source of historical information about the spiritualist movement as related to the fourth dimension.

Schubert winds up his essay with these stirring words: "The high eminence on which the knowledge and civilization of humanity now stands was not reached by the thoughtless employment of fanciful ideas, nor by recourse to a four-dimensional world, but by hard, serious labor, and slow, unceasing research. Let all men of science, therefore, band themselves together and oppose a solid front to methods that explain everything that is now mysterious to us by the interference of independent spirits."

W. Whately Smith, *A Theory of the Mechanism of Survival: The Fourth Dimension and its Applications* (Dutton & Co., New York, 1920).
The early part of our century marked a high point of popular interest in the four dimension. Spiritualism, with its 4-D spirits, was all the rage, and the Einstein-Minkowski use of the fourth dimension had given it a sort of legitimacy in the public mind.

Smith's book contains a nice Abbott-style description of the fourth dimension by analogy. Interestingly, he requires that the 2-D beings crawl back and forth around the 1-D rim of a 2-D disc, just as we 3-D beings are compelled to move back and forth on the 2-D surface of a 3-D sphere. He introduces the notion of time as a higher dimension in terms of making a stack of pictures taken of the 2-D world.

In the second part of the book he recounts various spiritualist experiences (the most compelling of which are the memories of a man who was revived from apparent death) and attempts to tie these experiences into the notion of the fourth dimension. Smith's idea is that one's consciousness has a 4-D "vehicle" as well as the familiar 3-D one. This modeling, however, is not carried out at any level much more convincing than the observation that the fourth dimension and spiritualist phenomena are both somehow transcendent.

One of the book's most memorable phrases arises when Smith describes, as an example of an out-and-out hallucination, the case of a man who enters his sitting room and finds "three green cassowaries playing nap."

Edwin F. Taylor and John A. Wheeler, *Spacetime Physics* (W. H. Freeman, San Francisco, 1963).
If my book has made you want to learn more about the special theory of

relativity, then this is the best place to go. Originally developed for the beginning of a Freshman Physics course, *Spacetime Physics* will not soon be replaced. The book's supple style and wealth of figures and tables make it a pleasure to read, and the 90 pages of thoroughly explained exercises encourage the reader to develop a real mastery of the material. There is also an elegant chapter on the Einstein theory of gravitation.

Bob Toben, Jack Sarfatti and Fred Wolf, *Space-Time and Beyond* (E. P. Dutton & Co., New York, 1975).

This glossy paperback consists of some 120 pages of arch drawings and hand-printed slogans by Bob Toben, followed by a short "scientific commentary" by Jack Sarfatti.

Some really interesting topics (e.g., Wheeler's quantum foam, sub-atomic black and white holes, and parallel universes) are discussed, but the authors seem determined to convince the reader of the validity of psychic phenomena as well. The scientific ideas are more often *invoked* than *explained*, and one leaves the book with little more than the impression that anything goes.

Johann Carl Friederich Zöllner, *Transcendental Physics*.

This strange book is the account of an astronomer's adventures with a spiritualist medium called Slade. As I mentioned in Chapter 1, Slade persuaded Zöllner that he was in contact with 4-D spirits in a number of ways, e.g., by getting them to write a message on a slate that had been sealed in a box with a bit of chalk. Slade, however, failed whenever he was confronted with a specific challenge, e.g., to turn the crystals of a chemical compound into crystals of the compound whose molecules were mirror images of the original compound.

But Zöllner's enthusiasm was so great that these repeated failures never seemed to shake him. Even if Slade could not do what he had been asked to, he would always come up with something. For instance, when asked to turn a seashell into its mirror image, Slade made it "pass through the table top" instead.

All this is perhaps reminiscent of Uri Geller's recent demonstrations of his psychic powers to certain interested scientists. Incidentally, I first heard of Zöllner in Martin Gardner's interesting book, *The Ambidextrous Universe*. In Gardner's book one can also find references to descriptions of Slade's eventual loss of credibility.

A CATALOGUE OF SELECTED DOVER BOOKS
IN ALL FIELDS OF INTEREST

A CATALOGUE OF SELECTED DOVER
BOOKS IN ALL FIELDS OF INTEREST

CELESTIAL OBJECTS FOR COMMON TELESCOPES, T. W. Webb. The most used book in amateur astronomy: inestimable aid for locating and identifying nearly 4,000 celestial objects. Edited, updated by Margaret W. Mayall. 77 illustrations. Total of 645pp. 5⅜ x 8½.
20917-2, 20918-0 Pa., Two-vol. set $9.00

HISTORICAL STUDIES IN THE LANGUAGE OF CHEMISTRY, M. P. Crosland. The important part language has played in the development of chemistry from the symbolism of alchemy to the adoption of systematic nomenclature in 1892. ". . . wholeheartedly recommended,"—Science. 15 illustrations. 416pp. of text. 5⅝ x 8¼.
63702-6 Pa. $6.00

BURNHAM'S CELESTIAL HANDBOOK, Robert Burnham, Jr. Thorough, readable guide to the stars beyond our solar system. Exhaustive treatment, fully illustrated. Breakdown is alphabetical by constellation: Andromeda to Cetus in Vol. 1; Chamaeleon to Orion in Vol. 2; and Pavo to Vulpecula in Vol. 3. Hundreds of illustrations. Total of about 2000pp. 6⅛ x 9¼.
23567-X, 23568-8, 23673-0 Pa., Three-vol. set $26.85

THEORY OF WING SECTIONS: INCLUDING A SUMMARY OF AIR-FOIL DATA, Ira H. Abbott and A. E. von Doenhoff. Concise compilation of subatomic aerodynamic characteristics of modern NASA wing sections, plus description of theory. 350pp. of tables. 693pp. 5⅜ x 8½.
60586-8 Pa. $7.00

DE RE METALLICA, Georgius Agricola. Translated by Herbert C. Hoover and Lou H. Hoover. The famous Hoover translation of greatest treatise on technological chemistry, engineering, geology, mining of early modern times (1556). All 289 original woodcuts. 638pp. 6¾ x 11.
60006-8 Clothbd. $17.95

THE ORIGIN OF CONTINENTS AND OCEANS, Alfred Wegener. One of the most influential, most controversial books in science, the classic statement for continental drift. Full 1966 translation of Wegener's final (1929) version. 64 illustrations. 246pp. 5⅜ x 8½. 61708-4 Pa. $4.50

THE PRINCIPLES OF PSYCHOLOGY, William James. Famous long course complete, unabridged. Stream of thought, time perception, memory, experimental methods; great work decades ahead of its time. Still valid, useful; read in many classes. 94 figures. Total of 1391pp. 5⅜ x 8½.
20381-6, 20382-4 Pa., Two-vol. set $13.00

UNCLE SILAS, J. Sheridan LeFanu. Victorian Gothic mystery novel, considered by many best of period, even better than Collins or Dickens. Wonderful psychological terror. Introduction by Frederick Shroyer. 436pp. 5⅜ x 8½.   21715-9 Pa. $6.00

JURGEN, James Branch Cabell. The great erotic fantasy of the 1920's that delighted thousands, shocked thousands more. Full final text, Lane edition with 13 plates by Frank Pape. 346pp. 5⅜ x 8½.
23507-6 Pa. $4.50

THE CLAVERINGS, Anthony Trollope. Major novel, chronicling aspects of British Victorian society, personalities. Reprint of Cornhill serialization, 16 plates by M. Edwards; first reprint of full text. Introduction by Norman Donaldson. 412pp. 5⅜ x 8½.   23464-9 Pa. $5.00

KEPT IN THE DARK, Anthony Trollope. Unusual short novel about Victorian morality and abnormal psychology by the great English author. Probably the first American publication. Frontispiece by Sir John Millais. 92pp. 6½ x 9¼.   23609-9 Pa. $2.50

RALPH THE HEIR, Anthony Trollope. Forgotten tale of illegitimacy, inheritance. Master novel of Trollope's later years. Victorian country estates, clubs, Parliament, fox hunting, world of fully realized characters. Reprint of 1871 edition. 12 illustrations by F. A. Faser. 434pp. of text. 5⅜ x 8½.   23642-0 Pa. $5.00

YEKL and THE IMPORTED BRIDEGROOM AND OTHER STORIES OF THE NEW YORK GHETTO, Abraham Cahan. Film *Hester Street* based on *Yekl* (1896). Novel, other stories among first about Jewish immigrants of N.Y.'s East Side. Highly praised by W. D. Howells—Cahan "a new star of realism." New introduction by Bernard G. Richards. 240pp. 5⅜ x 8½.   22427-9 Pa. $3.50

THE HIGH PLACE, James Branch Cabell. Great fantasy writer's enchanting comedy of disenchantment set in 18th-century France. Considered by some critics to be even better than his famous *Jurgen*. 10 illustrations and numerous vignettes by noted fantasy artist Frank C. Pape. 320pp. 5⅜ x 8½.   23670-6 Pa. $4.00

ALICE'S ADVENTURES UNDER GROUND, Lewis Carroll. Facsimile of ms. Carroll gave Alice Liddell in 1864. Different in many ways from final Alice. Handlettered, illustrated by Carroll. Introduction by Martin Gardner. 128pp. 5⅜ x 8½.   21482-6 Pa. $2.00

FAVORITE ANDREW LANG FAIRY TALE BOOKS IN MANY COLORS, Andrew Lang. The four Lang favorites in a boxed set—the complete *Red, Green, Yellow* and *Blue* Fairy Books. 164 stories; 439 illustrations by Lancelot Speed, Henry Ford and G. P. Jacomb Hood. Total of about 1500pp. 5⅜ x 8½.   23407-X Boxed set, Pa. $14.95

TONE POEMS, SERIES II: TILL EULENSPIEGELS LUSTIGE STREICHE, ALSO SPRACH ZARATHUSTRA, AND EIN HELDEN-LEBEN, Richard Strauss. Three important orchestral works, including very popular *Till Eulenspiegel's Marry Pranks,* reproduced in full score from original editions. Study score. 315pp. 9⅜ x 12¼. (Available in U.S. only)
23755-9 Pa. $7.50

TONE POEMS, SERIES I: DON JUAN, TOD UND VERKLARUNG AND DON QUIXOTE, Richard Strauss. Three of the most often performed and recorded works in entire orchestral repertoire, reproduced in full score from original editions. Study score. 286pp. 9⅜ x 12¼. (Available in U.S. only)
23754-0 Pa. $7.50

11 LATE STRING QUARTETS, Franz Joseph Haydn. The form which Haydn defined and "brought to perfection." (*Grove's*). 11 string quartets in complete score, his last and his best. The first in a projected series of the complete Haydn string quartets. Reliable modern Eulenberg edition, otherwise difficult to obtain. 320pp. 8⅜ x 11¼. (Available in U.S. only)
23753-2 Pa. $6.95

FOURTH, FIFTH AND SIXTH SYMPHONIES IN FULL SCORE, Peter Ilyitch Tchaikovsky. Complete orchestral scores of Symphony No. 4 in F Minor, Op. 36; Symphony No. 5 in E Minor, Op. 64; Symphony No. 6 in B Minor, "Pathetique," Op. 74. Bretikopf & Hartel eds. Study score. 480pp. 9⅜ x 12¼.
23861-X Pa. $10.95

THE MARRIAGE OF FIGARO: COMPLETE SCORE, Wolfgang A. Mozart. Finest comic opera ever written. Full score, not to be confused with piano renderings. Peters edition. Study score. 448pp. 9⅜ x 12¼. (Available in U.S. only)
23751-6 Pa. $11.95

"IMAGE" ON THE ART AND EVOLUTION OF THE FILM, edited by Marshall Deutelbaum. Pioneering book brings together for first time 38 groundbreaking articles on early silent films from *Image* and 263 illustrations newly shot from rare prints in the collection of the International Museum of Photography. A landmark work. Index. 256pp. 8¼ x 11.
23777-X Pa. $8.95

AROUND-THE-WORLD COOKY BOOK, Lois Lintner Sumption and Marguerite Lintner Ashbrook. 373 cooky and frosting recipes from 28 countries (America, Austria, China, Russia, Italy, etc.) include Viennese kisses, rice wafers, London strips, lady fingers, hony, sugar spice, maple cookies, etc. Clear instructions. All tested. 38 drawings. 182pp. 5⅜ x 8.
23802-4 Pa. $2.50

THE ART NOUVEAU STYLE, edited by Roberta Waddell. 579 rare photographs, not available elsewhere, of works in jewelry, metalwork, glass, ceramics, textiles, architecture and furniture by 175 artists—Mucha, Seguy, Lalique, Tiffany, Gaudin, Hohlwein, Saarinen, and many others. 288pp. 8⅜ x 11¼.
23515-7 Pa. $6.95

YUCATAN BEFORE AND AFTER THE CONQUEST, Diego de Landa. First English translation of basic book in Maya studies, the only significant account of Yucatan written in the early post-Conquest era. Translated by distinguished Maya scholar William Gates. Appendices, introduction, 4 maps and over 120 illustrations added by translator. 162pp. 5⅜ x 8½.
23622-6 Pa. $3.00

THE MALAY ARCHIPELAGO, Alfred R. Wallace. Spirited travel account by one of founders of modern biology. Touches on zoology, botany, ethnography, geography, and geology. 62 illustrations, maps. 515pp. 5⅜ x 8½.
20187-2 Pa. $6.95

THE DISCOVERY OF THE TOMB OF TUTANKHAMEN, Howard Carter, A. C. Mace. Accompany Carter in the thrill of discovery, as ruined passage suddenly reveals unique, untouched, fabulously rich tomb. Fascinating account, with 106 illustrations. New introduction by J. M. White. Total of 382pp. 5⅜ x 8½. (Available in U.S. only) 23500-9 Pa. $4.00

THE WORLD'S GREATEST SPEECHES, edited by Lewis Copeland and Lawrence W. Lamm. Vast collection of 278 speeches from Greeks up to present. Powerful and effective models; unique look at history. Revised to 1970. Indices. 842pp. 5⅜ x 8½. 20468-5 Pa. $8.95

THE 100 GREATEST ADVERTISEMENTS, Julian Watkins. The priceless ingredient; His master's voice; 99 44/100% pure; over 100 others. How they were written, their impact, etc. Remarkable record. 130 illustrations. 233pp. 7⅞ x 10 3/5. 20540-1 Pa. $5.00

CRUICKSHANK PRINTS FOR HAND COLORING, George Cruickshank. 18 illustrations, one side of a page, on fine-quality paper suitable for watercolors. Caricatures of people in society (c. 1820) full of trenchant wit. Very large format. 32pp. 11 x 16. 23684-6 Pa. $5.00

THIRTY-TWO COLOR POSTCARDS OF TWENTIETH-CENTURY AMERICAN ART, Whitney Museum of American Art. Reproduced in full color in postcard form are 31 art works and one shot of the museum. Calder, Hopper, Rauschenberg, others. Detachable. 16pp. 8¼ x 11.
23629-3 Pa. $2.50

MUSIC OF THE SPHERES: THE MATERIAL UNIVERSE FROM ATOM TO QUASAR SIMPLY EXPLAINED, Guy Murchie. Planets, stars, geology, atoms, radiation, relativity, quantum theory, light, antimatter, similar topics. 319 figures. 664pp. 5⅜ x 8½.
21809-0, 21810-4 Pa., Two-vol. set $10.00

EINSTEIN'S THEORY OF RELATIVITY, Max Born. Finest semi-technical account; covers Einstein, Lorentz, Minkowski, and others, with much detail, much explanation of ideas and math not readily available elsewhere on this level. For student, non-specialist. 376pp. 5⅜ x 8½.
60769-0 Pa. $4.50

CATALOGUE OF DOVER BOOKS

THE AMERICAN SENATOR, Anthony Trollope. Little known, long un-available Trollope novel on a grand scale. Here are humorous comment on American vs. English culture, and stunning portrayal of a heroine/villainess. Superb evocation of Victorian village life. 561pp. 5⅜ x 8½.
23801-6 Pa. $6.00

WAS IT MURDER? James Hilton. The author of *Lost Horizon* and *Goodbye, Mr. Chips* wrote one detective novel (under a pen-name) which was quickly forgotten and virtually lost, even at the height of Hilton's fame. This edition brings it back—a finely crafted public school puzzle resplendent with Hilton's stylish atmosphere. A thoroughly English thriller by the creator of Shangri-la. 252pp. 5⅜ x 8. (Available in U.S. only)
23774-5 Pa. $3.00

CENTRAL PARK: A PHOTOGRAPHIC GUIDE, Victor Laredo and Henry Hope Reed. 121 superb photographs show dramatic views of Central Park: Bethesda Fountain, Cleopatra's Needle, Sheep Meadow, the Blockhouse, plus people engaged in many park activities: ice skating, bike riding, etc. Captions by former Curator of Central Park, Henry Hope Reed, provide historical view, changes, etc. Also photos of N.Y. landmarks on park's periphery. 96pp. 8½ x 11.
23750-8 Pa. $4.50

NANTUCKET IN THE NINETEENTH CENTURY, Clay Lancaster. 180 rare photographs, stereographs, maps, drawings and floor plans recreate unique American island society. Authentic scenes of shipwreck, lighthouses, streets, homes are arranged in geographic sequence to provide walking-tour guide to old Nantucket existing today. Introduction, captions. 160pp. 8⅞ x 11¾.
23747-8 Pa. $6.95

STONE AND MAN: A PHOTOGRAPHIC EXPLORATION, Andreas Feininger. 106 photographs by *Life* photographer Feininger portray man's deep passion for stone through the ages. Stonehenge-like megaliths, fortified towns, sculpted marble and crumbling tenements show textures, beauties, fascination. 128pp. 9¼ x 10¾.
23756-7 Pa. $5.95

CIRCLES, A MATHEMATICAL VIEW, D. Pedoe. Fundamental aspects of college geometry, non-Euclidean geometry, and other branches of mathematics: representing circle by point. Poincare model, isoperimetric property, etc. Stimulating recreational reading. 66 figures. 96pp. 5⅝ x 8¼.
63698-4 Pa. $2.75

THE DISCOVERY OF NEPTUNE, Morton Grosser. Dramatic scientific history of the investigations leading up to the actual discovery of the eighth planet of our solar system. Lucid, well-researched book by well-known historian of science. 172pp. 5⅜ x 8½.
23726-5 Pa. $3.00

THE DEVIL'S DICTIONARY. Ambrose Bierce. Barbed, bitter, brilliant witticisms in the form of a dictionary. Best, most ferocious satire America has produced. 145pp. 5⅜ x 8½.
20487-1 Pa. $2.00

SECOND PIATIGORSKY CUP, edited by Isaac Kashdan. One of the greatest tournament books ever produced in the English language. All 90 games of the 1966 tournament, annotated by players, most annotated by both players. Features Petrosian, Spassky, Fischer, Larsen, six others. 228pp. 5⅜ x 8½. 23572-6 Pa. $3.50

ENCYCLOPEDIA OF CARD TRICKS, revised and edited by Jean Hugard. How to perform over 600 card tricks, devised by the world's greatest magicians: impromptus, spelling tricks, key cards, using special packs, much, much more. Additional chapter on card technique. 66 illustrations. 402pp. 5⅜ x 8½. (Available in U.S. only) 21252-1 Pa. $3.95

MAGIC: STAGE ILLUSIONS, SPECIAL EFFECTS AND TRICK PHO-TOGRAPHY, Albert A. Hopkins, Henry R. Evans. One of the great classics; fullest, most authorative explanation of vanishing lady, levitations, scores of other great stage effects. Also small magic, automata, stunts. 446 illus-trations. 556pp. 5⅜ x 8½. 23344-8 Pa. $6.95

THE SECRETS OF HOUDINI, J. C. Cannell. Classic study of Houdini's incredible magic, exposing closely-kept professional secrets and revealing, in general terms, the whole art of stage magic. 67 illustrations. 279pp. 5⅜ x 8½. 22913-0 Pa. $3.00

HOFFMANN'S MODERN MAGIC, Professor Hoffmann. One of the best, and best-known, magicians' manuals of the past century. Hundreds of tricks from card tricks and simple sleight of hand to elaborate illusions involving construction of complicated machinery. 332 illustrations. 563pp. 5⅜ x 8½. 23623-4 Pa. $6.00

MADAME PRUNIER'S FISH COOKERY BOOK, Mme. S. B. Prunier. More than 1000 recipes from world famous Prunier's of Paris and London, specially adapted here for American kitchen. Grilled tournedos with anchovy butter, Lobster a la Bordelaise, Prunier's prized desserts, more. Glossary. 340pp. 5⅜ x 8½. (Available in U.S. only) 22679-4 Pa. $3.00

FRENCH COUNTRY COOKING FOR AMERICANS, Louis Diat. 500 easy-to-make, authentic provincial recipes compiled by former head chef at New York's Fitz-Carlton Hotel: onion soup, lamb stew, potato pie, more. 309pp. 5⅜ x 8½. 23665-X Pa. $3.95

SAUCES, FRENCH AND FAMOUS, Louis Diat. Complete book gives over 200 specific recipes: bechamel, Bordelaise, hollandaise, Cumberland, apri-cot, etc. Author was one of this century's finest chefs, originator of vichyssoise and many other dishes. Index. 156pp. 5⅜ x 8. 23663-3 Pa. $2.50

TOLL HOUSE TRIED AND TRUE RECIPES, Ruth Graves Wakefield. Authentic recipes from the famous Mass. restaurant: popovers, veal and ham loaf, Toll House baked beans, chocolate cake crumb pudding, much more. Many helpful hints. Nearly 700 recipes. Index. 376pp. 5⅜ x 8½. 23560-2 Pa. $4.50

PRINCIPLES OF ORCHESTRATION, Nikolay Rimsky-Korsakov. Great classical orchestrator provides fundamentals of tonal resonance, progression of parts, voice and orchestra, tutti effects, much else in major document. 330pp. of musical excerpts. 489pp. 6½ x 9¼.   21266-1 Pa. $6.00

TRISTAN UND ISOLDE, Richard Wagner. Full orchestral score with complete instrumentation. Do not confuse with piano reduction. Commentary by Felix Mottl, great Wagnerian conductor and scholar. Study score. 655pp. 8⅛ x 11.   22915-7 Pa. $12.50

REQUIEM IN FULL SCORE, Giuseppe Verdi. Immensely popular with choral groups and music lovers. Republication of edition published by C. F. Peters, Leipzig, n. d. German frontmaker in English translation. Glossary. Text in Latin. Study score. 204pp. 9⅜ x 12¼.
                                                                23682-X Pa. $6.00

COMPLETE CHAMBER MUSIC FOR STRINGS, Felix Mendelssohn. All of Mendelssohn's chamber music: Octet, 2 Quintets, 6 Quartets, and Four Pieces for String Quartet. (Nothing with piano is included). Complete works edition (1874-7). Study score. 283 pp. 9⅜ x 12¼.
                                                                23679-X Pa. $6.95

POPULAR SONGS OF NINETEENTH-CENTURY AMERICA, edited by Richard Jackson. 64 most important songs: "Old Oaken Bucket," "Arkansas Traveler," "Yellow Rose of Texas," etc. Authentic original sheet music, full introduction and commentaries. 290pp. 9 x 12.   23270-0 Pa. $6.00

COLLECTED PIANO WORKS, Scott Joplin. Edited by Vera Brodsky Lawrence. Practically all of Joplin's piano works—rags, two-steps, marches, waltzes, etc., 51 works in all. Extensive introduction by Rudi Blesh. Total of 345pp. 9 x 12.   23106-2 Pa. $14.95

BASIC PRINCIPLES OF CLASSICAL BALLET, Agrippina Vaganova. Great Russian theoretician, teacher explains methods for teaching classical ballet; incorporates best from French, Italian, Russian schools. 118 illustrations. 175pp. 5⅜ x 8½.   22036-2 Pa. $2.50

CHINESE CHARACTERS, L. Wieger. Rich analysis of 2300 characters according to traditional systems into primitives. Historical-semantic analysis to phonetics (Classical Mandarin) and radicals. 820pp. 6⅛ x 9¼.
                                                                21321-8 Pa. $10.00

EGYPTIAN LANGUAGE: EASY LESSONS IN EGYPTIAN HIERO-GLYPHICS, E. A. Wallis Budge. Foremost Egyptologist offers Egyptian grammar, explanation of hieroglyphics, many reading texts, dictionary of symbols. 246pp. 5 x 7½. (Available in U.S. only)
                                                                21394-3 Clothbd. $7.50

AN ETYMOLOGICAL DICTIONARY OF MODERN ENGLISH, Ernest Weekley. Richest, fullest work, by foremost British lexicographer. Detailed word histories. Inexhaustible. Do not confuse this with *Concise Etymological Dictionary*, which is abridged. Total of 856pp. 6½ x 9¼.
                                          21873-2, 21874-0 Pa., Two-vol. set $12.00

"OSCAR" OF·THE WALDORF'S COOKBOOK, Oscar Tschirky. Famous American chef reveals 3455 recipes that made Waldorf great; cream of French, German, American cooking, in all categories. Full instructions, easy home use. 1896 edition. 907pp. 6⅝ x 9⅜.   20790-0 Clothbd. $15.00

COOKING WITH BEER, Carole Fahy. Beer has as superb an effect on food as wine, and at fraction of cost. Over 250 recipes for appetizers, soups, main dishes, desserts, breads, etc. Index. 144pp. 5⅜ x 8½. (Available in U.S. only)   23661-7 Pa. $2.50

STEWS AND RAGOUTS, Kay Shaw Nelson. This international cookbook offers wide range of 108 recipes perfect for everyday, special occasions, meals-in-themselves, main dishes. Economical, nutritious, easy-to-prepare: goulash, Irish stew, boeuf bourguignon, etc. Index. 134pp. 5⅜ x 8½.
                                                             23662-5 Pa. $2.50

DELICIOUS MAIN COURSE DISHES, Marian Tracy. Main courses are the most important part of any meal. These 200 nutritious, economical recipes from around the world make every meal a delight. "I . . . have found it so useful in my own household,"—N.Y. Times. Index. 219pp. 5⅜ x 8½.   23664-1 Pa. $3.00

FIVE ACRES AND INDEPENDENCE, Maurice G. Kains. Great back-to-the-land classic explains basics of self-sufficient farming: economics, plants, crops, animals, orchards, soils, land selection, host of other necessary things. Do not confuse with skimpy faddist literature; Kains was one of America's greatest agriculturalists. 95 illustrations. 397pp. 5⅜ x 8½.
                                                             20974-1 Pa. $3.95

A PRACTICAL GUIDE FOR THE BEGINNING FARMER, Herbert Jacobs. Basic, extremely useful first book for anyone thinking about moving to the country and starting a farm. Simpler than Kains, with greater emphasis on country living in general. 246pp. 5⅜ x 8½.
                                                             23675-7 Pa. $3.50

A GARDEN OF PLEASANT FLOWERS (PARADISI IN SOLE: PARADISUS TERRESTRIS), John Parkinson. Complete, unabridged reprint of first (1629) edition of earliest great English book on gardens and gardening. More than 1000 plants & flowers of Elizabethan, Jacobean garden fully described, most with woodcut illustrations. Botanically very reliable, a "speaking garden" of exceeding charm. 812 illustrations. 628pp. 8½ x 12¼.   23392-8 Clothbd. $25.00

ACKERMANN'S COSTUME PLATES, Rudolph Ackermann. Selection of 96 plates from the Repository of Arts, best published source of costume for English fashion during the early 19th century. 12 plates also in color. Captions, glossary and introduction by editor Stella Blum. Total of 120pp. 8⅜ x 11¼.   23690-0 Pa. $4.50

HOUSEHOLD STORIES BY THE BROTHERS GRIMM. All the great Grimm stories: "Rumpelstiltskin," "Snow White," "Hansel and Gretel," etc., with 114 illustrations by Walter Crane. 269pp. 5⅜ x 8½.
21080-4 Pa. $3.00

SLEEPING BEAUTY, illustrated by Arthur Rackham. Perhaps the fullest, most delightful version ever, told by C. S. Evans. Rackham's best work. 49 illustrations. 110pp. 7⅞ x 10¾.
22756-1 Pa. $2.50

AMERICAN FAIRY TALES, L. Frank Baum. Young cowboy lassoes Father Time; dummy in Mr. Floman's department store window comes to life; and 10 other fairy tales. 41 illustrations by N. P. Hall, Harry Kennedy, Ike Morgan, and Ralph Gardner. 209pp. 5⅜ x 8½.
23643-9 Pa. $3.00

THE WONDERFUL WIZARD OF OZ, L. Frank Baum. Facsimile in full color of America's finest children's classic. Introduction by Martin Gardner. 143 illustrations by W. W. Denslow. 267pp. 5⅜ x 8½.
20691-2 Pa. $3.50

THE TALE OF PETER RABBIT, Beatrix Potter. The inimitable Peter's terrifying adventure in Mr. McGregor's garden, with all 27 wonderful, full-color Potter illustrations. 55pp. 4¼ x 5½. (Available in U.S. only)
22827-4 Pa. $1.25

THE STORY OF KING ARTHUR AND HIS KNIGHTS, Howard Pyle. Finest children's version of life of King Arthur. 48 illustrations by Pyle. 131pp. 6⅛ x 9¼.
21445-1 Pa. $4.95

CARUSO'S CARICATURES, Enrico Caruso. Great tenor's remarkable caricatures of self, fellow musicians, composers, others. Toscanini, Puccini, Farrar, etc. Impish, cutting, insightful. 473 illustrations. Preface by M. Sisca. 217pp. 8⅜ x 11¼.
23528-9 Pa. $6.95

PERSONAL NARRATIVE OF A PILGRIMAGE TO ALMADINAH AND MECCAH, Richard Burton. Great travel classic by remarkably colorful personality. Burton, disguised as a Moroccan, visited sacred shrines of Islam, narrowly escaping death. Wonderful observations of Islamic life, customs, personalities. 47 illustrations. Total of 959pp. 5⅜ x 8½.
21217-3, 21218-1 Pa., Two-vol. set $12.00

INCIDENTS OF TRAVEL IN YUCATAN, John L. Stephens. Classic (1843) exploration of jungles of Yucatan, looking for evidences of Maya civilization. Travel adventures, Mexican and Indian culture, etc. Total of 669pp. 5⅜ x 8½.
20926-1, 20927-X Pa., Two-vol. set $7.90

AMERICAN LITERARY AUTOGRAPHS FROM WASHINGTON IRVING TO HENRY JAMES, Herbert Cahoon, et al. Letters, poems, manuscripts of Hawthorne, Thoreau, Twain, Alcott, Whitman, 67 other prominent American authors. Reproductions, full transcripts and commentary. Plus checklist of all American Literary Autographs in The Pierpont Morgan Library. Printed on exceptionally high-quality paper. 136 illustrations. 212pp. 9⅛ x 12¼.
23548-3 Pa. $7.95

AMERICAN BIRD ENGRAVINGS, Alexander Wilson et al. All 76 plates. from Wilson's *American Ornithology* (1808-14), most important ornithological work before Audubon, plus 27 plates from the supplement (1825-33) by Charles Bonaparte. Over 250 birds portrayed. 8 plates also reproduced in full color. 111pp. 9⅜ x 12½. 23195-X Pa. $6.00

CRUICKSHANK'S PHOTOGRAPHS OF BIRDS OF AMERICA, Allan D. Cruickshank. Great ornithologist, photographer presents 177 closeups, groupings, panoramas, flightings, etc., of about 150 different birds. Expanded *Wings in the Wilderness*. Introduction by Helen G. Cruickshank. 191pp. 8¼ x 11. 23497-5 Pa. $6.00

AMERICAN WILDLIFE AND PLANTS, A. C. Martin, et al. Describes food habits of more than 1000 species of mammals, birds, fish. Special treatment of important food plants. Over 300 illustrations. 500pp. 5⅜ x 8½. 20793-5 Pa. $4.95

THE PEOPLE CALLED SHAKERS, Edward D. Andrews. Lifetime of research, definitive study of Shakers: origins, beliefs, practices, dances, social organization, furniture and crafts, impact on 19th-century USA, present heritage. Indispensable to student of American history, collector. 33 illustrations. 351pp. 5⅜ x 8½. 21081-2 Pa. $4.00

OLD NEW YORK IN EARLY PHOTOGRAPHS, Mary Black. New York City as it was in 1853-1901, through 196 wonderful photographs from N.-Y. Historical Society. Great Blizzard, Lincoln's funeral procession, great buildings. 228pp. 9 x 12. 22907-6 Pa. $7.95

MR. LINCOLN'S CAMERA MAN: MATHEW BRADY, Roy Meredith. Over 300 Brady photos reproduced directly from original negatives, photos. Jackson, Webster, Grant, Lee, Carnegie, Barnum; Lincoln; Battle Smoke, Death of Rebel Sniper, Atlanta Just After Capture. Lively commentary. 368pp. 8⅜ x 11¼. 23021-X Pa. $8.95

TRAVELS OF WILLIAM BARTRAM, William Bartram. From 1773-8, Bartram explored Northern Florida, Georgia, Carolinas, and reported on wild life, plants, Indians, early settlers. Basic account for period, entertaining reading. Edited by Mark Van Doren. 13 illustrations. 141pp. 5⅜ x 8½. 20013-2 Pa. $4.50

THE GENTLEMAN AND CABINET MAKER'S DIRECTOR, Thomas Chippendale. Full reprint, 1762 style book, most influential of all time; chairs, tables, sofas, mirrors, cabinets, etc. 200 plates, plus 24 photographs of surviving pieces. 249pp. 9⅞ x 12¾. 21601-2 Pa. $6.50

AMERICAN CARRIAGES, SLEIGHS, SULKIES AND CARTS, edited by Don H. Berkebile. 168 Victorian illustrations from catalogues, trade journals, fully captioned. Useful for artists. Author is Assoc. Curator, Div. of Transportation of Smithsonian Institution. 168pp. 8½ x 9½. 23328-6 Pa. $5.00

THE COMPLETE BOOK OF DOLL MAKING AND COLLECTING, Catherine Christopher. Instructions, patterns for dozens of dolls, from rag doll on up to elaborate, historically accurate figures. Mould faces, sew clothing, make doll houses, etc. Also collecting information. Many illustrations. 288pp. 6 x 9. 22066-4 Pa. $4.50

THE DAGUERREOTYPE IN AMERICA, Beaumont Newhall. Wonderful portraits, 1850's townscapes, landscapes; full text plus 104 photographs. The basic book. Enlarged 1976 edition. 272pp. 8¼ x 11¼. 23322-7 Pa. $7.95

CRAFTSMAN HOMES, Gustav Stickley. 296 architectural drawings, floor plans, and photographs illustrate 40 different kinds of "Mission-style" homes from The Craftsman (1901-16), voice of American style of simplicity and organic harmony. Thorough coverage of Craftsman idea in text and picture, now collector's item. 224pp. 8⅛ x 11. 23791-5 Pa. $6.00

PEWTER-WORKING: INSTRUCTIONS AND PROJECTS, Burl N. Osborn. & Gordon O. Wilber. Introduction to pewter-working for amateur craftsman. History and characteristics of pewter; tools, materials, step-by-step instructions. Photos, line drawings, diagrams. Total of 160pp. 7⅞ x 10¾. 23786-9 Pa. $3.50

THE GREAT CHICAGO FIRE, edited by David Lowe. 10 dramatic, eye-witness accounts of the 1871 disaster, including one of the aftermath and rebuilding, plus 70 contemporary photographs and illustrations of the ruins—courthouse, Palmer House, Great Central Depot, etc. Introduction by David Lowe. 87pp. 8¼ x 11. 23771-0 Pa. $4.00

SILHOUETTES: A PICTORIAL ARCHIVE OF VARIED ILLUSTRA-TIONS, edited by Carol Belanger Grafton. Over 600 silhouettes from the 18th to 20th centuries include profiles and full figures of men and women, children, birds and animals, groups and scenes, nature, ships, an alphabet. Dozens of uses for commercial artists and craftspeople. 144pp. 8⅜ x 11¼. 23781-8 Pa. $4.00

ANIMALS: 1,419 COPYRIGHT-FREE ILLUSTRATIONS OF MAM-MALS, BIRDS, FISH, INSECTS, ETC., edited by Jim Harter. Clear wood engravings present, in extremely lifelike poses, over 1,000 species of animals. One of the most extensive copyright-free pictorial sourcebooks of its kind. Captions. Index. 284pp. 9 x 12. 23766-4 Pa. $7.95

INDIAN DESIGNS FROM ANCIENT ECUADOR, Frederick W. Shaffer. 282 original designs by pre-Columbian Indians of Ecuador (500-1500 A.D.). Designs include people, mammals, birds, reptiles, fish, plants, heads, geometric designs. Use as is or alter for advertising, textiles, leathercraft, etc. Introduction. 95pp. 8¾ x 11¼. 23764-8 Pa. $3.50

SZIGETI ON THE VIOLIN, Joseph Szigeti. Genial, loosely structured tour by premier violinist, featuring a pleasant mixture of reminiscences, insights into great music and musicians, innumerable tips for practicing violinists. 385 musical passages. 256pp. 5⅝ x 8¼. 23763-X Pa. $3.50

CATALOGUE OF DOVER BOOKS

THE STANDARD BOOK OF QUILT MAKING AND COLLECTING, Marguerite Ickis. Full information, full-sized patterns for making 46 traditional quilts, also 150 other patterns. Quilted cloths, lame, satin quilts, etc. 483 illustrations. 273pp. 6⅞ x 9⅝.   20582-7 Pa. $4.95

ENCYCLOPEDIA OF VICTORIAN NEEDLEWORK, S. Caulfield, Blanche Saward. Simply inexhaustible gigantic alphabetical coverage of every traditional needlecraft—stitches, materials, methods, tools, types of work; definitions, many projects to be made. 1200 illustrations; double-columned text. 697pp. 8⅛ x 11.   22800-2, 22801-0 Pa., Two-vol. set $12.00

MECHANICK EXERCISES ON THE WHOLE ART OF PRINTING, Joseph Moxon. First complete book (1683-4) ever written about typography, a compendium of everything known about printing at the latter part of 17th century. Reprint of 2nd (1962) Oxford Univ. Press edition. 74 illustrations. Total of 550pp. 6⅛ x 9¼.   23617-X Pa. $7.95

PAPERMAKING, Dard Hunter. Definitive book on the subject by the foremost authority in the field. Chapters dealing with every aspect of history of craft in every part of the world. Over 320 illustrations. 2nd, revised and enlarged (1947) edition. 672pp. 5⅜ x 8½.   23619-6 Pa. $7.95

THE ART DECO STYLE, edited by Theodore Menten. Furniture, jewelry, metalwork, ceramics, fabrics, lighting fixtures, interior decors, exteriors, graphics from pure French sources. Best sampling around. Over 400 photographs. 183pp. 8⅜ x 11¼.   22824-X Pa. $6.00

Prices subject to change without notice.

Available at your book dealer or write for free catalogue to Dept. GI, Dover Publications, Inc., 180 Varick St., N.Y., N.Y. 10014. Dover publishes more than 175 books each year on science, elementary and advanced mathematics, biology, music, art, literary history, social sciences and other areas.